Capability Planning and Analysis to Optimize **Air Force Intelligence, Surveillance, and Reconnaissance Investments**

Committee on Examination of the Air Force Intelligence, Surveillance, and Reconnaissance (ISR) Capability Planning and Analysis (CP&A) Process

Air Force Studies Board

Division on Engineering and Physical Sciences

NATIONAL RESEARCH COUNCIL
OF THE NATIONAL ACADEMIES

THE NATIONAL ACADEMIES PRESS
Washington, D.C.
www.nap.edu

THE NATIONAL ACADEMIES PRESS 500 Fifth Street, NW Washington, DC 20001

NOTICE: The project that is the subject of this report was approved by the Governing Board of the National Research Council, whose members are drawn from the councils of the National Academy of Sciences, the National Academy of Engineering, and the Institute of Medicine. The members of the committee responsible for the report were chosen for their special competences and with regard for appropriate balance.

This is a report of work supported by Grant FA 9550-11-1-0126 between the U.S. Air Force and the National Academy of Sciences. Any opinions, findings, conclusions, or recommendations expressed in this publication are those of the author(s) and do not necessarily reflect the view of the organizations or agencies that provided support for the project.

International Standard Book Number-13: 978-0-309-25814-2
International Standard Book Number-10: 0-309-25814-6

Limited copies of this report are available from:

Air Force Studies Board
National Research Council
500 Fifth Street, NW
Washington, DC 20001
(202) 334-3111

Additional copies are available from:

The National Academies Press
500 Fifth Street, NW
Keck 360
Washington, DC 20001
(800) 624-6242 or (202) 334-3313
http://www.nap.edu

Printed in the United States of America

THE NATIONAL ACADEMIES

Advisers to the Nation on Science, Engineering, and Medicine

The **National Academy of Sciences** is a private, nonprofit, self-perpetuating society of distinguished scholars engaged in scientific and engineering research, dedicated to the furtherance of science and technology and to their use for the general welfare. Upon the authority of the charter granted to it by the Congress in 1863, the Academy has a mandate that requires it to advise the federal government on scientific and technical matters. Dr. Ralph J. Cicerone is president of the National Academy of Sciences.

The **National Academy of Engineering** was established in 1964, under the charter of the National Academy of Sciences, as a parallel organization of outstanding engineers. It is autonomous in its administration and in the selection of its members, sharing with the National Academy of Sciences the responsibility for advising the federal government. The National Academy of Engineering also sponsors engineering programs aimed at meeting national needs, encourages education and research, and recognizes the superior achievements of engineers. Dr. Charles M. Vest is president of the National Academy of Engineering.

The **Institute of Medicine** was established in 1970 by the National Academy of Sciences to secure the services of eminent members of appropriate professions in the examination of policy matters pertaining to the health of the public. The Institute acts under the responsibility given to the National Academy of Sciences by its congressional charter to be an adviser to the federal government and, upon its own initiative, to identify issues of medical care, research, and education. Dr. Harvey V. Fineberg is president of the Institute of Medicine.

The **National Research Council** was organized by the National Academy of Sciences in 1916 to associate the broad community of science and technology with the Academy's purposes of furthering knowledge and advising the federal government. Functioning in accordance with general policies determined by the Academy, the Council has become the principal operating agency of both the National Academy of Sciences and the National Academy of Engineering in providing services to the government, the public, and the scientific and engineering communities. The Council is administered jointly by both Academies and the Institute of Medicine. Dr. Ralph J. Cicerone and Dr. Charles M. Vest are chair and vice chair, respectively, of the National Research Council.

www.national-academies.org

COMMITTEE ON EXAMINATION OF THE AIR FORCE INTELLIGENCE, SURVEILLANCE, AND RECONNAISSANCE (ISR) CAPABILITY PLANNING AND ANALYSIS (CP&A) PROCESS

BRIAN A. ARNOLD, Raytheon Company, *Co-Chair*
LAWRENCE J. DELANEY, Titan Corporation (retired), *Co-Chair*
COLLIN A. AGEE, U.S. Army
MELANI AUSTIN, Lockheed Martin Aeronautics Company
THOMAS J. BURNS, Science Applications International Corporation
PAMELA A. DREW, TASC
RAND H. FISHER, The Aerospace Corporation
KEITH R. HALL, Booz Allen Hamilton
LESLIE F. KENNE, LK Associates
ROBERT H. LATIFF, R. Latiff Associates
TERRY P. LEWIS, Raytheon Company
MICHAEL A. LONGORIA, The RAND Corporation
PAUL F. McMANAMON, Exciting Technology, LLC
MATT L. MLEZIVA, Wildwood Strategic Concepts, LLC
GERALD F. PERRYMAN, JR., Independent Consultant
JONATHAN M. SMITH, University of Pennsylvania

Staff

CARTER W. FORD, Study Director
GREGORY EYRING, Senior Program Officer
SARAH M. CAPOTE, Research Associate
ZEIDA PATMON, Program Associate

Preface

Prior to 2009 the U.S. Air Force did not have a comprehensive approach for investing in and acquiring intelligence, surveillance, and reconnaissance (ISR) capabilities. In 2009, the Air Force developed and implemented the ISR Flight Plan to focus Air Force needs on future ISR capabilities and has subsequently renamed this approach Capability Planning and Analysis (CP&A), which shares characteristics of, but does not equate to, the Air Force development planning process.[1] In 2011, the Air Force requested that the National Research Council (NRC), under the auspices of the Air Force Studies Board (AFSB), undertake a study to improve this process, specifically to provide the Air Force foundational analytics to aid decision making, especially in light of overall future defense spending. In response to this request, the NRC established the Committee on Examination of the Air Force ISR CP&A Process. Biographical information for the committee members is provided in Appendix A. The terms of reference for the study are presented in Box 1-1 in Chapter 1.

The AFSB was established in 1996 as a unit of the NRC at the request of the U.S. Air Force. The AFSB brings to bear broad military, industrial, and academic scientific, engineering, and management expertise on Air Force technical challenges and other issues of importance to senior Air Force leaders. The board discusses potential studies of interest, develops and frames study tasks, ensures proper project planning, suggests potential committee members and reviewers for reports produced by fully independent ad hoc study committees, and convenes meetings to examine strategic issues. The board members were not asked to endorse the committee's conclusions or recommendations, nor did they review the final draft of this report

[1]U.S. Air Force. 2010. *Development Planning Guide*. Wright-Patterson Air Force Base: Air Force Materiel Command Directorate of Intelligence and Requirements. June.

before its release, although board members with appropriate expertise may be nominated to serve as formal members of study committees or as report reviewers.

The committee thanks the many people who provided it with information for the study, including the guest speakers shown in Appendix B, their organizations, and supporting staff members; and others, including the study sponsors Dr. Steven Walker, Deputy Assistant Secretary of the Air Force for Science, Technology, and Engineering, and Lt Gen Larry James, Deputy Chief of Staff for ISR, and their staff members.

Brian A. Arnold, *Co-Chair*
Lawrence J. Delaney, *Co-Chair*
Committee on Examination of the Air Force Intelligence, Surveillance, and Reconnaissance (ISR) Capability Planning and Analysis (CP&A) Process

Acknowledgment of Reviewers

This report has been reviewed in draft form by individuals chosen for their diverse perspectives and technical expertise, in accordance with procedures approved by the National Research Council's Report Review Committee. The purpose of this independent review is to provide candid and critical comments that will assist the institution in making its published report as sound as possible and to ensure that the report meets institutional standards for objectivity, evidence, and responsiveness to the study charge. The review comments and draft manuscript remain confidential to protect the integrity of the deliberative process. We wish to thank the following individuals for their review of this report:

William P. Delaney, Massachusetts Institute of Technology Lincoln Laboratory,
Ronald P. Fuchs, Independent Consultant,
Richard L. Garwin, IBM Thomas J. Watson Research Center,
Mark Lewis, IDA Science and Technology Policy Institute,
Anthony Metoyer, The Boeing Company,
Thomas E. Romesser, Northrop Grumman Corporation (retired),
Peter B. Teets, U.S. Air Force/National Reconnaissance Office (retired), and
Alan R. Washburn, U.S. Naval Postgraduate School.

Although the reviewers listed above provided many constructive comments and suggestions, they were not asked to endorse the conclusions or recommendations, nor did they see the final draft of the report before its release. The review of this report was overseen by Julia M. Phillips, Sandia National Laboratories, and Robert J. Hermann, U.S. Air Force/National Reconnaissance Office (retired). Appointed by the National Research Council, they were responsible for making certain that an independent examination of this report was carried out in accordance with institutional procedures and that all review comments were carefully considered. Responsibility for the final content of this report rests entirely with the authoring committee and the institution.

Contents

SUMMARY 1

1 INTELLIGENCE, SURVEILLANCE, AND RECONNAISSANCE CHALLENGES FACING THE AIR FORCE 9

Introduction, 9
Committee Formation and Terms of Reference, 11
Study Approach, 12
Scenarios That May Guide Air Force ISR Force-Planning Processes, 13
Regionally Specific (Traditional) Scenarios, 14
Global (Non-Traditional) Scenarios, 20
Homeland Security-Based Scenarios, 21
Organization of the Report, 23

2 THE CURRENT STATE OF THE AIR FORCE INTELLIGENCE, SURVEILLANCE, AND RECONNAISSANCE INVESTMENT PLANNING PROCESS 24

Introduction, 24
Historical Development of the Air Force ISR Planning Process, 29
The ISR Flight Plan, 31
The Core Function Lead Integrator Construct, 33
The Current Air Force ISR Capability Planning and Analysis Process, 34
The Core Function Lead Integrator Process, 36
Space Superiority and Cyberspace Superiority Core Function Master Plans, 39

Integration of Air Force Core Function Master Plans, 41
ISR Capability Planning and Analysis and Core Function Master Plan Link with Planning, Programming, Budgeting, and Execution, 41
Linkages Between the Air Force and the Intelligence Community, 42
Findings, 43
Concluding Thoughts, 45

3 EXAMPLES OF PROCESSES EMPLOYED BY GOVERNMENT AND INDUSTRY FOR PROVIDING CAPABILITY PLANNING AND ANALYSIS 47
Introduction, 47
Examples of Government Processes for Providing Capability Planning and Analysis, 47
U.S. Army, 48
U.S. Navy, 52
Office of the Under Secretary of Defense for Intelligence, 55
Examples of Industry Processes for Providing Capability Planning and Analysis, 59
Booz Allen Hamilton, 59
TASC, 61
RadiantBlue, Inc., 67
Concluding Thoughts, 72

4 TOWARD AN ENHANCED AIR FORCE INTELLIGENCE, SURVEILLANCE, AND RECONNAISSANCE CAPABILITY PLANNING AND ANALYSIS PROCESS 74
Introduction, 74
Recommendations, 74
Desired Attributes of an Enterprise-Wide Intelligence, Surveillance, and Reconnaissance Capability Planning and Analysis Process, 77
Proposed Air Force Intelligence, Surveillance, and Reconnaissance Capability Planning and Analysis Process, 80
Problem Definition and Approach, 80
Needs Analysis, 84
Multi-resolution Gap Analysis, 86
Solution Analysis, 91
Concluding Thoughts, 94

APPENDIXES

A Biographical Sketches of Committee Members 97
B List of Committee Meetings, Presenters, and Participating Organizations 106
C Supplement to Chapter 3: Descriptions of Additional Organizational CP&A Processes and Tools 113

Acronyms

A2/AD	anti-access/area denial
ACC	Air Combat Command
ACL	Achievable Capabilities List
AF	Air Force
AF/A2	Deputy Chief of Staff of the Air Force for ISR
AFCS	Air Force Corporate Structure
AFI	Air Force Instruction
AFISRA	Air Force ISR Agency
AFMC	Air Force Materiel Command
AFROC	Air Force Requirements Oversight Council
AFSPC	Air Force Space Command
AOA	analysis of alternatives
AOC	Air Operations Center
APPG	annual planning and programming guidance
ARFORGEN	Army Force Generation
BA	Battlespace Awareness
BA CIB	Battlespace Awareness Capabilities Integration Board
BAH	Booz Allen Hamilton
BAPA	Battlespace Awareness and Portfolio Assessment
BCT	Brigade Combat Teams
BES	Budget Estimate Submission
BMDS	Ballistic Missile Defense System

C2	command and control
C4I	command, control, communications, computers, and intelligence
CA&P	capability assessment and planning
CADD	Capability Area Deep Dive
CAPE	Cost Assessment and Program Evaluation
CART	Capabilities Analysis Requirements Tool
CASA	Communications Architecture Systems Assessor
CBA	Capabilities-Based Assessment
CBP	capability-based planning
CBPfM	Capabilities-Based Portfolio Management
CDD	Capability Description Document
CDRUSSTRATCOM	Commander, U.S. Strategic Command
CET	Capabilities Effectiveness Tool
CFLI	Core Function Lead Integrator
CFMP	Core Function Master Plan
CJCS	Chairman of the Joint Chiefs of Staff
CJCSI	Chairman of the Joint Chiefs of Staff Instruction
CNO	Chief of Naval Operations
COCOM	Combatant Command
COIN	counterinsurgency
COMINT	communications intelligence
CONOPS	concept of operations
CoP	community of practice
COTS	commercial off-the-shelf
CP&A	Capability Planning and Analysis
CPM	Capability Portfolio Manager
CRD	Capabilities Requirements Document
CRRA	Capability Review and Risk Assessment
CSA	Coalition Situational Awareness
CSAF	Chief of Staff of the Air Force
DAWG	Deputy's Advisory Working Group
DCGS	Distributed Common Ground Station
DCR	DOTMLPF Change Recommendation
DMZ	Demilitarized Zone
DNI	Director of National Intelligence
DoD	Department of Defense
DOTMLPF	Doctrine, Organization, Training, Materiel, Leadership and Education, Personnel and Facilities

DOTMLPF-P	Doctrine, Organization, Training, Materiel, Leadership and Education, Personnel, Facilities, and Policy
DOT_LPF	non-material analysis
DPRK	Democratic People's Republic of Korea
DRU	Direct Reporting Unit
E/CCA	Element/Component Characterizations for Analysis
EEI	essential elements of information
EFDS	Expeditionary Force Development System
ELINT	electronics intelligence
EMD	Engineering and Manufacturing Development
EO	electro-optical (imaging)
ESA	electronically scanned array
FCB	Functional Capabilities Board
FOA	Field Operating Agency
FY	Fiscal Year
FYDP	Future Years Defense Program
GAO	Government Accountability Office
GIISR	Global Integrated ISR
GMTI	ground moving target indicator
GOTS	government off-the-shelf
HAF	Headquarters (U.S.) Air Force
HLS	Homeland Security
HIS	hyperspectral imaging
HUMINT	human intelligence
IC	intelligence community
ICD	Initial Capabilities Document
IED	improvised explosive device
IPL	integrated priority list
IR	infrared
IROC	Intelligence Readiness Operations Capability
ISAR	Inverse Synthetic Aperture Radar
ISCA	Integrated Sensor Coverage Area
ISR	intelligence, surveillance, and reconnaissance
ISR-CART	ISR Capabilities Analysis Requirements Tool
ITW	integrated tactical warning

JBA	Joint Battlespace Awareness
JCA	Joint Capability Area
JCIDS	Joint Capabilities Integration and Development System
JFC	Joint Functional Concept
JFCC	Joint Functional Component Command
JOC	Joint Operations Center; Joint Operating Concept
JROC	Joint Requirements Oversight Council
JTF	Joint Task Force
JUON	Joint Urgent Operational Need
JWICS	Joint Worldwide Intelligence Communications System
M&S	modeling and simulation
MAJCOM	Major Command
MCO	Major Combat Operations
MDA	Milestone Decision Authority; Missile Defense Agency
MGA	Multi-resolution Gap Analysis
MI	Military Intelligence
MIP	Military Intelligence Program
MO	Mission Overwatch
MOE	measures of effectiveness
MOP	measures of performance
MOU	measures of utility
MRA	Multi-Resolution Analysis
MSA	modeling, simulation, and analysis
MSI	multispectral imager
MTI	Moving Target Indicator
NCDP	Naval Capabilities Development Process
NGA	National Geospatial-Intelligence Agency
NGC	Northrop Grumman Corporation
NIIRS	National Imagery Interpretability Rating Scale
NIP	National Intelligence Program
NMS	National Military Strategy
NRC	National Research Council
NRO	National Reconnaissance Office
NSA	National Security Agency
NTISR	non-traditional ISR
OCO	Overseas Contingency Operation(s)
ORS	Operationally Responsive Space
OSD	Office of the Secretary of Defense

OUSD(AT&L)	Office of the Under Secretary of Defense for Acquisition, Technology, and Logistics
OUSD(I)	Office of the Under Secretary of Defense for Intelligence
PAA	Persistent Area Assessment
PCL	Prioritized Capability List
PCPAD	planning and direction, collection, processing and exploitation, analysis and production, and dissemination
PDA	Problem Definition and Approach
PED	processing, exploitation, and dissemination
Pk	probability of kill
PLA	People's Liberation Army
POC	point of contact
POM	Program Objective Memorandum
PoR	Program of Record
PP&R	Portfolios, Programs, and Resources
PPBES	planning, programming, budgeting, and execution system
QDR	Quadrennial Defense Review
QRC	Quick Reaction Capabilities
R&D	research and development
RCS	radar cross section
RMD	resource management decision
ROK	Republic of Korea
RPA	remotely piloted aircraft
SAR	synthetic aperture radar
SCADA	supervisory control and data acquisition
SCF	Service Core Function
SEA	Strategic Environmental Assessment
SEAS	System Effectiveness Analysis Simulation
SECAF	Secretary of the Air Force
SECDEF	Secretary of Defense
SETA	Systems Engineering and Technical Assistance
SID	Situation Development
SIGINT	signals intelligence
SIPRnet	Secret Internet Protocol Router Network
SLRG	Senior Level Review Group

SMC/XR	Space and Missile Systems Center, Directorate of Development Planning
SME	subject matter expert
SOAP	Satellite Orbit Analysis Program
SSDR	security system dynamically reconfigurable
STK	Satellite Tool Kit®
SYSSIM	System Simulation
TASC	The Analytical Sciences Corporation
TCPED	Tasking, Collection, Processing, Exploitation, and Dissemination
TOA	Total Obligation Authority
TOR	terms of reference
TPED	tasking, processing, exploitation, and dissemination
UGS	unattended ground sensor
UON	urgent operational need
USAF	United States Air Force
USCYBERCOM	U.S. Cyber Command
USN	United States Navy
USSTRATCOM	U.S. Strategic Command
VCJCS	Vice Chairman of the Joint Chiefs of Staff
VCSAF	Vice Chief of Staff of the Air Force
WIP	Warfighter Involvement Process
WMD	weapons of mass destruction
Wx	weather

Summary

CURRENT CONTEXT

Intelligence, surveillance, and reconnaissance (ISR) capabilities have expanded situation awareness for U.S. forces, provided for more precise combat effects, and enabled better decision making both during conflicts and in peacetime, and reliance on ISR capabilities is expected to increase in the future. ISR capabilities are critical to 3 of the 12 Service Core Functions of the U.S. Air Force (USAF): namely, Global Integrated ISR (GIISR) and the ISR components of Cyberspace Superiority and Space Superiority, and they contribute to all others.[1,2,3] The rapid growth and

[1]"ISR" is defined as "[a]n activity that synchronizes and integrates the planning and operation of sensors, assets, and processing, exploitation, and dissemination systems in direct support of current and future operations. This is an integrated intelligence and operations function." SOURCE: Department of Defense (DoD). 2010. "Department of Defense Dictionary of Military and Associated Terms (Joint Publication 1-02). 8 November. As amended through 15 October 2011." Available at http://www.dtic.mil/doctrine/new_pubs/jp1_02.pdf. Accessed February 6, 2012.

[2]"Service Core Functions define the Air Force's key capabilities and contributions as a service. Service Core Functions correspond to the specific primary functions of the service as described in DoD Directive 5100.01." SOURCE: USAF. 2012. "GIISR Operations. Air Force Doctrine Document 2-0." January 6.

[3]Following are the names of the Air Force Service Core Functions: (1) Nuclear Deterrence Operations, (2) Air Superiority, (3) Global Precision Attack, (4) Personnel Recovery, (5) Command and Control, (6) Global Integrated ISR, (7) Space Superiority, (8) Cyberspace Superiority, (9) Rapid Global Mobility, (10) Special Operations, (11) Agile Combat Support, and (12) Building Partnerships. SOURCE: Col Brian Johnson, Chief, ISR Plans and Integration Division (AF/A2DP), Headquarters, U.S. Air Force. "Air Force ISR: CP&A Overview." Presentation to the committee, October 6, 2011.

evolution of the use of Air Force ISR capabilities since September 11, 2001, have been focused largely on immediate requirements dictated by the wars in Afghanistan and Iraq. Managing this enterprise intelligently has involved many challenges, including the following: (1) the diverse mission and information requirements in the military services and the intelligence community (IC)[4]; (2) the diverse domains in which ISR operates (space, air, ground, sea, undersea, and cyberspace); (3) the need to balance joint versus organic ISR assets, and command and control; (4) the need to balance rapid-acquisition capabilities that will satisfy urgent warfighter needs versus capabilities that will satisfy long-term strategic goals; and (5) the need to balance sensor data-collection capability against capabilities for planning and direction, collection, processing and exploitation, analysis and production, and dissemination (PCPAD).

Recognizing these challenges, the Air Force undertook a series of organizational changes, beginning in 2006 with the establishment of the flag officer position of Deputy Chief of Staff of the Air Force for ISR (AF/A2), followed in 2007 with the creation of the Air Force ISR Agency.[5] In 2009, the Air Force developed and implemented the ISR Flight Plan process to focus Air Force needs on future ISR capabilities.[6] The Air Force subsequently renamed this approach the Capability Planning and Analysis (CP&A) process "to align with [the] CFLI [Core Function Lead Integrator] construct."[7] The ISR CP&A process employs subject-matter experts from across the service who consider strategic guidance, analyze operational needs, determine operational gaps, conduct risk and solutions analysis, and produce a master plan to guide investment. The processes used are lengthy and personnel-intensive and cannot quickly respond to revisions in assumptions and requirements. There is considerable reason for and need to improve the present processes, especially to account for new ISR needs in the cyberspace and space domains.

In response to a request from AF/A2 and the Deputy Assistant Secretary of the Air Force for Science, Technology, and Engineering, the National Research Council (NRC), under the auspices of the Air Force Studies Board, formed the Committee on Examination of the Air Force Intelligence, Surveillance, and Reconnaissance

---

[4]The IC is composed of 17 member organizations and includes the National Reconnaissance Office, the National Security Agency, the Defense Intelligence Agency, and the National Geospatial-Intelligence Agency. For more information, see http://www.intelligence.gov/about-the-intelligence-community/member-agencies/. Accessed May 24, 2012.

[5]Col Brian Johnson, Chief, ISR Plans and Integration Division (AF/A2DP), Headquarters, U.S. Air Force. "Air Force ISR CP&A Overview." Presentation to the committee, October 6, 2011.

[6]Lt Gen David Deptula (USAF, Ret.), Chief Executive Officer and Managing Director, Mav6. "The Air Force ISR Flight Plan: Origin, Rational and Process." Presentation to the committee, October 6, 2011.

[7]Col Brian Johnson, Chief, ISR Plans and Integration Division (AF/A2DP), Headquarters, U.S. Air Force. "Air Force ISR CP&A Overview." Presentation to the committee, October 6, 2011.

(ISR) Capability Planning and Analysis (CP&A) Process. The terms of reference (TOR) for this study are as follows:

- Review the current approach to the Air Force corporate planning and programming process for ISR capability generation.
- Review various analytical methods, processes and models for large scale, complex domains like ISR and identify best practices.
- Apply the current approach and recommended best practices to the Air Force corporate planning and programming process for ISR, in the context of the future Joint, National, and Coalition partner environment.
- Recommend improvements/changes to existing analytical tools, methods, roles/responsibilities, organization and management, etc. that would be required to ensure that the Air Force corporate planning and programming process for ISR is successful in addressing all Joint, National, and Coalition partners' needs.[8]

In the double-numbering of the findings and recommendations presented in the next two sections, the first number reflects the chapter from which each is drawn. All 14 report findings and 3 report recommendations are presented in the Summary. Chapter 1 provides a broad context of historical factors related to the development of ISR capabilities and considers potential scenarios involving the use of these capabilities. Chapter 2 addresses Task 1 of the TOR by reviewing the current approach to the Air Force corporate planning and programming processes for ISR. Chapter 3 covers Task 2 of the TOR by reviewing various analytical method(s), processes, and models for large-scale, complex domains like ISR, and identifies best practices. Chapter 4 responds to Tasks 3 and 4 of the TOR by offering recommendations for Air Force consideration to improve its ISR CP&A process and an ideal model of an Air Force "system-of-systems" evaluation process for ISR CP&A. Findings are embedded in the text of Chapters 2 and 3 after the supporting evidence is presented.

THE FINDINGS OF THE COMMITTEE REGARDING THE AIR FORCE ISR CAPABILITY PLANNING AND ANALYSIS PROCESS

Finding 2-1. The responsibility for evaluating and informing decisions about Air Force ISR capabilities is diffuse, overly personnel-intensive, and divided among many organizations, resulting in an excessively lengthy process. Spe-

[8]Air Force ISR investments include the air, space, and cyberspace domains, which, in turn, provide critical inputs into the ground and maritime ISR domains. The Air Force sponsor requested that the committee focus specifically on the air, space, and cyberspace domains for this report.

cifically, the respective roles and responsibilities of the AF/A2 and the GIISR CFLI are not well defined or well understood, and appear disconnected. Both the ISR CP&A and the CFLI processes have positive aspects, but the processes are immature and insufficiently integrated.

Finding 2-2. The Air Force ISR planning process lacks adequate process definition and formal interaction between the Space Superiority, Cyberspace Superiority, and GIISR CFLIs. It also does not rigorously integrate ISR contributions from other military services, the IC, and the Office of the Secretary of Defense. Consequently, the Air Force process does not yield ISR investment priorities across domains and security constructs. The Air Force needs increased awareness of what capabilities it provides, along with the IC and other services, to the Joint fight to reduce duplication of effort and funds expended.

Finding 2-3. Air Force platforms do not appear to be included in Air Force cyberspace-related planning processes, even though cyberspace vulnerabilities do exist onboard platforms and in the connectivity between them. Moreover, cyberspace functions can play a very positive role in support of ISR, and ISR systems can help support cyberspace functions. Additionally, the complexity of the multi-organizational relationships involved in current DoD and IC interactions leads to confusion in both execution and planning processes, particularly for cyber operations.

Finding 2-4. The Air Force lacks integrated modeling and simulation and analysis tools that provide traceability from requirements to capability and that conduct operationally relevant ISR trade-space analysis across the doctrine, organization, training, materiel, leadership and education, personnel, facilities, and policy (DOTMLPF-P) framework and within and across air, space, and cyberspace domains.

Finding 2-5. The Air Force corporate process "disassembles" the ISR portfolio planning analysis, classifies the elements into isolated, or stovepipe, function components, and then makes trade-offs and/or decisions without the ISR trade-space underpinnings.

Finding 2-6. The ISR CP&A process lacks the ability to respond in a timely way with appropriate fidelity to meet the increasing speed of technology development, operational requirements, and the required decrease in planning-cycle time, particularly in the cyberspace domain.

Finding 2-7. PCPAD is not adequately considered and prioritized by the ISR CP&A process.

Finding 2-8. The ISR CP&A process does not adequately consider affordability in capability trade-space analysis.

Finding 3-1. The U.S. Army's Integrated Sensor Coverage Area (ISCA) construct uses a process that links requirements analysis with force development and portfolio management in a way that helps synchronize planning and execution. Keys to this linkage are the ISCA analytical underpinnings and the methodology that enables sensor-platform aggregations. Additionally, the ISCA construct uses measured performance to inform acquisition decisions in a manner that lends transparency, responsiveness, and repeatability.

Finding 3-2. The U.S. Navy's capability-based process is collaborative across the Department of the Navy and is synchronized with the planning, programming, budgeting, and execution system and system acquisition life cycles. The process can be streamlined to address urgent needs. The process deals largely with naval requirements; utilizes existing PCPAD/TCPED (tasking, collection, processing, exploitation, and dissemination) architectures; and connects with other ISR enterprise providers through the Office of the Under Secretary of Defense for Intelligence (OUSD[I]).

Finding 3-3. The CP&A-like process employed by OUSD(I) addresses ISR enterprise concerns across the DoD and the IC and includes consideration of the capabilities of enterprise networks and PCPAD and TCPED. The OUSD(I) recognizes the need to improve the capability development process in the following ways: (1) by attaining better up-front fidelity on trade-offs involving cost and schedule and performance, (2) by providing more analytic rigor and risk/portfolio analysis, (3) by placing stronger emphasis on prioritizing requirements and capabilities, and (4) by strengthening the alignment of the acquisition process.

Finding 3-4. Booz Allen Hamilton's Capabilities-Based Portfolio Management process requires leadership engagement, diverse skill sets to analyze a portfolio, and stakeholder participation and transparency. The resulting assessments are repeatable and rigorous enough to enable long-term planning, yet agile enough to incorporate new scenarios, priorities, and missions. The process includes the modeling of extant TCPED and communications architectures, which yields more realistic estimates of cost and performance and risk. Although many results are scalable, any consideration of broader, more complex enterprises requires good analytical judgment for the development of the right approach.

Finding 3-5. TASC's capability-based assessment process employs Multi-Resolution Analysis (MRA), which in turn allows the complexity of ISR to be handled in a straightforward, transparent, tailorable, scalable, repeatable manner, incorporating a suite of tools that are optimized for a specific purpose. Such an approach can support a wide range of decisions and decision time lines.

Finding 3-6. RadiantBlue's modeling, simulation, and analysis capability focuses on the physics-based capability and architecture analysis and mission utility analysis found in MRA. The BlueSim tool, combined with RadiantBlue's methodology, has been used to successfully support trade-space studies of various ISR and PED architectures.

RECOMMENDATIONS OF THE COMMITTEE FOR THE IMPROVEMENT OF THE AIR FORCE ISR CAPABILITY PLANNING AND ANALYSIS PROCESS

Recommendation 4-1. The Air Force should adopt an ISR CP&A process that incorporates the following attributes:

- Encompasses all ISR missions;
- Addresses all ISR domains and sources, including non-traditional ISR;
- Includes all ISR assets in a sensor-to-user chain (e.g., PCPAD and communications);
- Collaborates with ISR-related entities;
- Provides traceability from process inputs to outputs;
- Is mission/scenario-based;
- Is repeatable and enduring;
- Supports trade-off analyses;
- Is scalable in size, time, and resolution; and
- Reduces labor and cost over time.[9]

Figure S-1 is the graphical depiction of the proposed ISR CP&A process.[10]

[9]The committee acknowledges that any process needs to accommodate the use of all levels of classified material in the analysis. However, security and time constraints precluded the committee from making recommendations for multi-level security analysis. Chapters 2 and 4 provide supporting discussions.

[10]Chapter 4 of the report provides detailed descriptions of each step in the proposed ISR CP&A process.

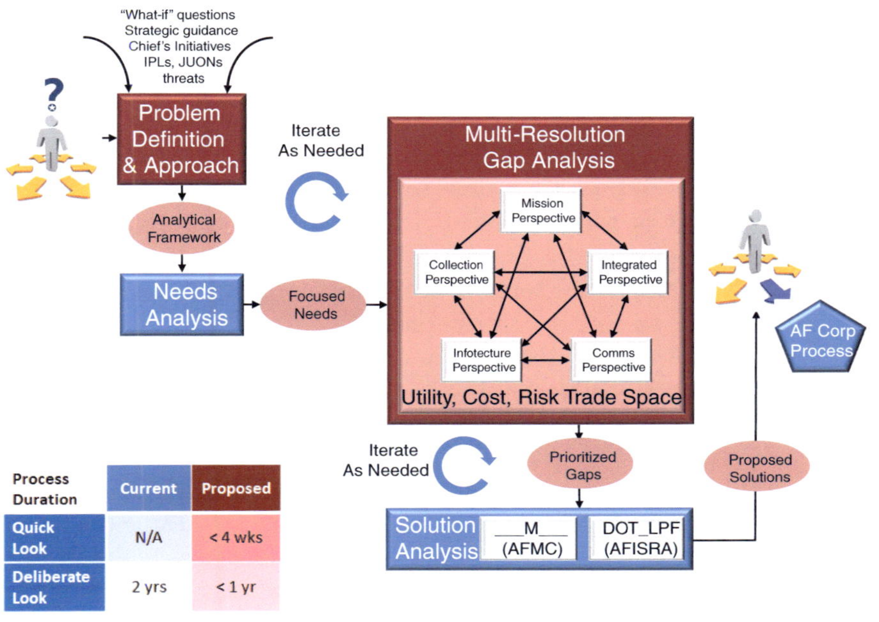

FIGURE S-1 A high-level diagram showing the major elements of the committee's proposed Air Force intelligence, surveillance, and reconnaissance Capability Planning and Analysis process. NOTE: Boxes and ovals shaded in red represent additions or modifications to the current process, depicted in blue. The table (lower left) indicates anticipated time lines for executing the process. The process is not intended to be strictly sequential in nature. Iterations may occur between various process functions as the analysis evolves. Acronyms are defined in the list in the front matter.

Recommendation 4-2. The Air Force should evolve its ISR CP&A process to an integrated, overarching ISR investment process with clear organizational responsibility identified for each subprocess.

Recommendation 4-3. The Air Force should adopt the proposed ISR CP&A process by incrementally building on its existing process using pilot projects. The scope of each pilot project should be compatible with available resources, be relevant to both current and future mission scenarios, and include metrics to measure achievement of the desired improvements (e.g., manpower reductions and increased timeliness).[11]

[11]The proposed process is described in Chapter 4. Also, notional scenarios are discussed in Chapter 1; they range from regional conflicts (Persian Gulf and Pacific Rim) to global, non-traditional conflicts, to homeland security scenarios.

WHY THE AIR FORCE SHOULD IMPLEMENT THE RECOMMENDATIONS FOR THE PROPOSED AIR FORCE ISR CAPABILITY PLANNING AND ANALYSIS PROCESS

Given the increasingly competitive, congested, contested, connected global environment, the U.S. military will continue to face numerous national security risks from a wide spectrum of real and potential adversaries. To address such risks, the DoD is increasingly encouraging closer working relationships between services and the IC in order to reduce redundancy of effort and funds expended. The Air Force also can improve its processes for contributing ISR capabilities to other services and the intelligence community.[12] The Chairman of the Joint Chiefs of Staff recently stated: "The U.S. armed services must achieve unprecedented synergy to ensure access to contested waters, skies, land, space and networks in the face of emerging weapons. . . ."[13] The importance of ISR systems in providing critical, essential, affordable contributions to our national security, including indications and warning, missile defense, and global strike, cannot be overstated. At the same time, there is a significant disconnect between those who view managing ISR as simply acquiring and managing more platforms and those who view managing ISR as acquiring and managing *capability. The value inherent in the proposed ISR CP&A process is sevenfold: (1) It enhances the quality, transparency, repeatability, and credibility of proposed investments. (2) It provides greater insight into cost, risk, and mission utility assessments. (3) It scales from quick-look through long-term analyses. (4) It expands the consideration and analysis of Joint and interagency capabilities. (5) It more fully addresses all ISR domains (air, space, land, maritime, cyberspace). (6) It encompasses the complete "sensor-to-user" chain including PCPAD. (7) It reduces the amount of time and labor required to answer investment questions.*

---

[12]The Government Accountability Office (GAO) noted in 2011: "The military services each have their own ISR plans and roadmaps that focus on their respective ISR activities and are not integrated with other services' plans. For example, the Air Force maintains its own ISR plan and metrics separate from DoD's ISR Integration Roadmap and the other service roadmaps, and the other services have developed several roadmaps outlining ISR priorities and capability gaps." SOURCE: GAO. 2011. "Intelligence, Surveillance, and Reconnaissance: Actions Are Needed to Increase Integration and Efficiencies of DoD's ISR Enterprise," p. 9. Available at http://www.gao.gov/assets/320/319163.pdf. Accessed March 21, 2012.

[13]Christopher J. Castelli. "Dempsey Urges Unprecedented Synergy to Counter New Threats," InsideDefense.com, December 7, 2011.

1

Intelligence, Surveillance, and Reconnaissance Challenges Facing the Air Force

INTRODUCTION

The United States is at a strategic turning point after a decade of war. As stated in the 2012 National Defense Strategy:

> Over the last decade, we have undertaken extended operations in Iraq and Afghanistan to bring stability to those countries and secure our interests. As we responsibly draw down from these two operations, take steps to protect our nation's economic vitality, and protect our interests in a world of accelerating change, we face an inflection point. . . . Out of the assessment we developed a defense strategy that transitions our Defense enterprise from an emphasis on today's wars to preparing for future challenges, protects the broad range of U.S. national security interests. . . .[1]

The evolving strategic landscape encompasses a vast list of uncertainties that include violent extremists, non-state actors, the proliferation of weapons of mass destruction, competition over dwindling natural resources, rapid growth in the availability and use of technology worldwide, growing global economic interdependency, and vulnerable and fragile commercial infrastructure.

For the foreseeable future, the United States will continue to take an active approach to countering threats related to the uncertainties listed above by monitoring

[1]Department of Defense (DoD). 2012. *Sustaining U.S. Global Leadership: Priorities for 21st Century Defense.* January. Available at http://www.defense.gov/news/Defense_Strategic_Guidance.pdf. Accessed February 29, 2012.

global activities.[2] The ability to carry out monitoring on a global scale will drive the importance of and dependence on intelligence, surveillance, and reconnaissance (ISR) for our nation in the future.[3] The range of ISR capabilities will expand to monitor terrorism, support irregular warfare, support power projection into anti-access/area denial (A2/AD) environments, monitor weapons of mass destruction and support arms control, defend our homeland, and provide support for response to natural disasters.[4]

Fiscal challenges, as always, will drive the need to allocate defense resources as efficiently as possible. This is especially true with respect to future ISR investments, because ISR touches all elements of the national security infrastructure as well as the nation's commercial infrastructure. Today's ISR capabilities consist of a mix of Cold War systems; modern air, space, and cyberspace systems; and a set of quick-reaction capabilities that were designed for specific point solutions. As the nation looks to the future, a key challenge will be how to integrate these existing capabilities with new capabilities to monitor the uncertain threats of the 21st century. The United States will continue to lead global efforts with capable allies and partners to ensure access to and use of the global commons. The fact that the United States operates in an integrated world and fights wars jointly and in coalitions drives the paramount need for coordinated and fully integrated ISR capabilities. The desired end state of a fully integrated ISR system drives the need for improved interoperability, commonality, and modernization overlaid on a set of standards, protocols, security, and open architectures.

Since September 2001, ISR capabilities have grown in importance and use by the Department of Defense (DoD) and the intelligence community (IC), in part because these capabilities provide information to the warfighter that serves as a force multiplier. This shared information enables better and faster decisions, precision effects, and lower risk for the commander in the field. Under the U.S. national security umbrella, the Air Force has a significant role in the acquisition, operation, and support of many ISR capabilities because it is simultaneously a user, a provider, and an operator in the Joint and coalition contexts. Air Force ISR capabilities deliver

---

[2]Although these threats need to be addressed by the Department of Defense, including all of the military services, the intelligence community, and the Department of Homeland Security, the focus of this research is directed particularly at Air Force intelligence, surveillance, and reconnaissance.

[3]"ISR" is defined as "[a]n activity that synchronizes and integrates the planning and operation of sensors, assets, and processing, exploitation, and dissemination systems in direct support of current and future operations. This is an integrated intelligence and operations function." SOURCE: DoD. 2010. "Department of Defense Dictionary of Military and Associated Terms (Joint Publication 1-02). 8 November. As amended through 15 October 2011." Available at http://www.dtic.mil/doctrine/new_pubs/jp1_02.pdf. Accessed February 6, 2012.

[4]DoD. 2012. *Sustaining U.S. Global Leadership: Priorities for 21st Century Defense.* January. Available at http://www.defense.gov/news/Defense_Strategic_Guidance.pdf. Accessed February 29, 2012.

needed information to strategic, operational, and tactical users alike, for operations from humanitarian assistance to active combat. As the demand for ISR capabilities grows, the Air Force experiences increasing pressure to allocate resources effectively and to acquire needed capabilities efficiently, on time, and on schedule. Additionally, Air Force ISR capabilities will be increasingly required to interoperate with capabilities managed by other U.S. organizations and coalition forces.

COMMITTEE FORMATION AND TERMS OF REFERENCE

In response to a request from the Deputy Assistant Secretary of the Air Force for Science, Technology, and Engineering, the National Research Council (NRC) formed the Committee on Examination of the Air Force Intelligence, Surveillance, and Reconnaissance (ISR) Capability Planning and Analysis (CP&A) Process. The NRC approved the terms of reference (TOR) for the study in March 2011 (see Box 1-1), and the Air Force funded this 18-month study in July 2011. Committee members were then selected and approved by the NRC for their backgrounds in academia, industry, and government (see Appendix A for biographical sketches of the committee members). Subject-matter support was provided by the Deputy Chief of Staff of the Air Force for ISR.

BOX 1-1
Terms of Reference

The NRC will:

1. Review the current approach to the Air Force corporate planning and programming process for ISR capability generation.
2. Review various analytical methods, processes and models for large scale, complex domains like ISR and identify best practices.
3. Apply the current approach and recommended best practices to the Air Force corporate planning and programming process for ISR, in the context of the future Joint, National and coalition partner environment.
4. Recommend improvements/changes to existing analytical tools, methods, roles/responsibilities, organization and management, etc. that would be required to ensure that the Air Force corporate planning and programming process for ISR is successful in addressing all Joint, National, and Coalition partners' needs.[a]

[a]Air Force ISR investments include the air, space, and cyberspace domains, which, in turn, provide critical inputs into the ground and maritime ISR domains. The Air Force sponsor requested that the committee focus specifically on the air, space, and cyberspace domains for this report.

The Air Force recognizes that an architectural perspective that includes the platforms, sensors, processors, terminals, and their connecting communications and data links—that is, an end-to-end solution—is a logical construct to drive the Air Force corporate planning and programming processes.

STUDY APPROACH

The committee held seven data-gathering meetings at which briefings were provided by senior leaders from the IC, which included DoD components and agencies (i.e., Air Force, Army, Navy, Office of the Secretary of Defense [OSD]), professional staff members from key congressional oversight committees, and senior industry executives (see Appendix B for a listing of meetings and participating organizations). In addition to its data-gathering sessions, the committee held two 3-day meetings to finalize its findings and recommendations. At all of these meetings, the committee discussed and evaluated what any proposed improvements to the Air Force CP&A process might accomplish, and it considered the following questions from the perspective of a high-level decision maker:

1. What capabilities do I need to acquire and when?
2. What capabilities should I retire and when?
3. On what analytical basis are my decisions made?
4. How much risk do I accept and when?
5. What are the level and range of uncertainty in my judgments?
6. Are there architectural or operational changes that could provide a dramatic, positive change in capability, and that would remain close to the current set of material solutions and/or cost?

To acquire the right capabilities, for the right reasons, under current and potential future circumstances, is extremely challenging.[5] Although the TOR for this study is specific to the Air Force, Air Force decisions about whether to enact the proposed ISR CP&A process will need to be made in the context of factors including but not limited to the following: (1) congressional support, (2) contract performance, (3) near-term versus far-term considerations and tactical versus strategic considerations, (4) requirements of other military services and the IC, and

[5] I.B. Holley. 1983. *Ideas and Weapons, Office of Air Force History Reprint.* Original Printing: Yale University Press, 1953: Introduction, p. v: "Since time immemorial weapons have played a significant role in tipping the scales of victory from one side to another . . . In recent years the pace has accelerated . . . the degree to which scientific and technological advances are exploited for military purposes depends upon the methods devised to that end. The haphazard and unsystematic means of other ages have yielded to a more orderly process of conscious decision, development, test, and evaluation, but even so these methods have lagged behind the creative forces of science."

(5) roles and responsibilities of the military services and the IC as defined by Title 10 versus Title 50 of the U.S. Code.[6] An ideal ISR CP&A process for the Air Force would provide answers to these questions and reasonably sustain decisions made over time in the context of the broad challenges of the 21st century.[7]

SCENARIOS THAT MAY GUIDE AIR FORCE ISR FORCE-PLANNING PROCESSES

Strategic requirements for the broad range of ISR capabilities are embedded within the 2012 National Defense Strategy.[8] Along with this new guidance, the DoD will base major force-planning efforts on a *prediction* of future conflicts and the *anticipated* requirements of existing and *"to be developed"* weapons systems. Budgetary restraint will also add significant risk that must be calculated into Air Force ISR force planning.[9] It is now assumed that the U.S. military will shrink over the next 10 years, through fiscal year (FY) 2022.[10]

It is anticipated that this new direction for the DoD will be realized by means of programmed budget reductions through the Future Years Defense Program and through FY 2022, similar to the post-war build-down after World War II, the

[6]The military services and the IC have specific roles and responsibilities, as defined by Title 10 and Title 50 of the *U.S. Code*, respectively. For additional information on Title 10, see http://uscode.house.gov/download/title_10.shtml. For additional information on Title 50, see http://uscode.house.gov/download/title_50.shtml. Accessed March 21, 2012.

[7]Chapter 4 provides a detailed discussion of the recommended ISR CP&A process.

[8]DoD. 2012. *Sustaining U.S. Global Leadership: Priorities for 21st Century Defense*. January, p. 1. Available at http://www.defense.gov/news/Defense_Strategic_Guidance.pdf. Accessed February 29, 2012.

[9]The Honorable Michael Donley, SECAF, and General Norton Schwartz, CSAF. Joint Statement, White Paper. "Air Force Priorities for a New Strategy with Constrained Budgets." February 2012. Available at http://www.af.mil/shared/media/document/AFD-120201-027.pdf. Accessed September 4, 2012. "Defense cuts totaling $487 billion over 10 years will be hard but manageable, though, significant challenges remain. The need to transition contingency appropriations into baseline budgets still overhangs DoD resource planning, excess basing capacity still needs to be addressed through the proposed Base Realignment and Closure Commission, and many more decisions due to unforeseen events will intervene in the next decade. The Air Force's FY13 budget request is the culmination of an unprecedented season of difficult choices. We can and expect to absorb currently programmed reductions with increased but acceptable risk, provided no further cuts are enacted. The possibility of the BCA reducing defense spending by billions more will put at risk our ability to execute the new strategic guidance."

[10]Craig Whitlock and Greg Jaffe. 2012. "Obama Announces New Leaner Military Approach." *Washington Post*, January 5. Available at http://www.washingtonpost.com/world/national-security/obama-announces-new-military-approach/2012/01/05/gIQAFWcmcP_story.html. Accessed April 12, 2012.

Korean War, and the Vietnam War.[11] In this larger context, the Air Force is developing a process to plan its ISR investment strategy that will likely be informed by the scenario-based modeling of potential future conflicts and the anticipated constraints on resources. The Air Force has used scenario-based planning, understanding its limitations, for over four decades.[12] The following sections briefly discuss several conflict scenarios in the context of regional, global, and homeland security challenges and venues in which the Air Force would apply various mixes of ISR capabilities. These sections are intended to provide the reader with a sense of the complexities involved in planning for future Air Force ISR capabilities—complexities that become even more complicated in a fiscally constrained environment.

Regionally Specific (Traditional) Scenarios

Overall, it is anticipated that combat operations will continue in Southwest Asia until 2014, with limited contingency and counterinsurgency (COIN) operations anticipated beyond the conclusion of major U.S. involvement in Afghanistan. DoD planners are also considering the potential for other future regional conflicts in addition to those in Southwest Asia. Any major regional conflict will require the operational surveillance of the entire regional or theater battlefield in order to underpin U.S. actions, even though the totality of U.S. military land forces may be dispersed into brigade- or regiment-sized elements focused by country, province, or village. Success in establishing persistent, theater-wide surveillance is normally considered by regional combatant commanders as the first priority, followed closely with as much COIN support as possible delivered directly to ground units. Representative Air Force ISR missions will include the following: theater-wide persistent situational awareness; high-value, time-critical targeting; countering of improvised explosive device; and COIN support for brigade- or regiment-sized ground forces arrayed across a region. In regional conflicts featuring significant U.S. ground force engagement that is not concentrated but distributed across an entire region or

[11]Gordon Adams. 2011. "Rethinking National Security in an Era of Declining Budgets." Johns Hopkins University/Applied Physics Laboratory 2011-2012. *Rethinking Seminar Series.* October 27.

[12]USAF Center for Strategy and Technology. Future Conflict Studies. Air University web site. "The premise of scenario planning is that it is better to get the future imprecisely right than to get the future precisely wrong. We know that our predictions of the future are never exactly correct. Rather than picking one definitive picture of the future and planning for that future, scenario planning allows a region to consider various possibilities and identify policies that can adapt to changing circumstances. Scenarios do not describe a forecasted end state. Scenarios are stories about future conditions that convey a range of possible outcomes." Available at http://csat.au.af.mil/future-conflict.htm#scenarios. Accessed March 28, 2012.

BOX 1-2
Regional Conflict Scenario:
Planning Assumptions for Air Force ISR Capabilities

1. The demand for localized, tactical-level surveillance will increase as more brigades deploy to various parts of a region.
2. Demands at a tactical level can rapidly exceed existing intelligence, surveillance, and reconnaissance (ISR) capabilities.
3. ISR priority systems must be very precise and highly flexible when meeting brigade and regiment demands, and if ground forces are not significantly involved, there will also be a need for ground truth.
4. Local understanding, cultural awareness, and anthropological depth may be the primary emphases in winning "hearts and minds" in a post-major-combat period, but ISR capabilities are required any time there are significant ground forces employed, whether they are engaged in intense combat or nation-building activities.[a]

[a]Noah Shachtman. 2012. "Air Force's Top Brain Wants a 'Social Radar' to 'See into Hearts and Minds,' " *Wired* Magazine Interview, January 19.

country, there are consistent planning assumptions about the Air Force ISR that will most likely withstand variances in any regional conflict model (see Box 1-2).

Even with the reduction in military forces in Iraq and Afghanistan, the Air Force will continue to plan for multiple scenarios that will involve various mixes of ISR capabilities.[13] There are many variations and permutations on predicting the immediate aspects of the national security environment,[14] but at least four other regional concern categories currently draw significant attention:

[13]Thomas Barnett. 2004. *The Pentagon's New Map: War and Peace in the Twenty-First Century*. New York: G.P. Putnam Sons.

[14]Andrew Krepinevich. 2009. *7 Deadly Scenarios: A Military Futurist Explores War in the 21st Century*. New York: Bantam Dell. The scenarios are as follows: (1) The collapse of Pakistan and the breakup of its army into loyalist and radical Islamist factions, both armed with nuclear weapons. (2) A series of terror attacks against cities in the United States involving stolen Russian nuclear warheads. (3) A new and deadly flu pandemic sweeping north into the United States from Mexico, causing massive refugee flows. (4) A new war against Israel by Hezbollah, with the backing of Iran. (5) Rising civil unrest in China, prompting the country to impose a blockade of Taiwan and threatening war against the United States if it intervenes. (6) A terrorist war on the global economy, by means of attacking infrastructure and logistics chains, and through sophisticated cyberattacks. (7) A civil war in Iraq following a dramatic reduction of U.S. troops.

1. Preventing Iranian nuclear development and aggression;
2. "Arab spring"[15] involvement in the Middle East;
3. Maintaining a balanced, mutually supportive relationship with China; and
4. Sustaining deterrence on the Korean Peninsula.

Persian Gulf

Increasingly, there is international realization that, even in the face of severe economic and political measures, the Iranian government is intent on developing nuclear weapons.[16] A confrontation with a nuclear-armed Iran has the potential to be a truly "hybrid" war, one that might require the U.S. military to counter Iran's conventional anti-access capabilities, defeat its irregular forces both at sea and on land, prepare for attacks by terrorist groups against American targets or U.S. allies globally, and, most importantly, conduct operations under the shadow of a possible nuclear attack.[17] In the event of a major confrontation, the United States would view this hybrid scenario in a regional war context, plus taking into account all of

[15]President Barack Obama. 2011. Speech and explanations from Ben Zimmer, *Visual Thesaurus*, "The Arab Spring Has Sprung," May 20, 2011: "*Arab spring* doesn't actually have to correspond to dates between the vernal equinox and the summer solstice, because *spring* is understood metaphorically and not literally. The obvious model for *Arab spring* is the *Prague spring* of 1968, when Czechoslovakia enjoyed a brief interval of democratic reform before the Soviet Union invaded. As Michael Quinion notes on his World Wide Words site, *Arab spring* and *Prague spring* have a much earlier precursor: the European revolutions of 1848, which historians dubbed *springtime of the peoples* or *spring of nations*. Those terms are translations of German *Völkerfrühling* and French *printemps des peuples*. From 1848 to 1968 to 2011, the social movements given the *spring* label have shared a hope for liberalization in the face of oppressive regimes." However, Barry Rubin (Director for Global Research in International Affairs) argues that, "in Middle Eastern usage it comes from the 'Beirut spring' in which hundreds of thousands of Lebanese demonstrated against the Syrian military presence and domination of the country. In the short term the Lebanese protesters won. But because of a lack of U.S. and Western help along with the ruthlessness of Syria, Iran, and their local allies (notably Hezbollah) the Beirut Spring . . . *was* defeated. Syria is back in control to a large degree and while the Syrian-backed government (including Hezbollah) has been kept at bay for months by bureaucratic maneuvers, presumably it will get into power at some point. So the term 'Arab Spring' is appropriate if we remember that the Beirut Spring, a good example of what's being faced now, turned into the Beirut Winter."

[16]Mark Gunzinger and Christopher Dougherty. 2012. *Outside-In: Operating from Range to Defeat Iran's Anti-Access and Area-Denial Threat,* January 17. Washington, D.C.: Center for Strategic and Budgetary Assessments.

[17]Ibid.

the other contingency models for worldwide counterterrorism, and possibly facing the complication of nuclear weapons.[18]

According to the Government Accountability Office (GAO): "In light of Iran's pursuit of A2/AD capabilities, it seems unlikely that the U.S. military's legacy planning assumptions will remain valid."[19] Accordingly, it is increasingly important for Air Force ISR force planners to look in detail at the A2/AD challenges as well as the exacting requirements for strategic and operational targeting.

Pacific Rim

In the Pacific Rim, Air Force planners may be considering multiple scenarios, ranging from no ongoing military conflict but in which some degree of military action can be foreseen in operations short of war, to operations that would presage conflict with the intended effect of deterring aggressive military action. Further, although that intention to deter military action may be steadfast, the possibility of escalation remains. The challenge in Southeast Asia with respect to North Korea and China is representative of this situation.

> Arguably the greatest strategic choice concerns how best to respond to China's rapid rise as a major power. Boasting the world's second-largest economy, Beijing has undertaken a decade-long military buildup of its People's Liberation Army (PLA). Its focus is on the Western Pacific, declared a vital interest by every U.S. administration for more than 60 years, with security commitments to such allies as Australia, Japan and South Korea, and states like Taiwan.[20]

China and the United States are linked by both economics and politics to the extent that it is in the best interests of both countries to maintain a stable Pacific region.

---

[18]The Government Accountability Office (GAO) noted: "[F]uture adversaries are likely to use 'hybrid warfare' tactics, a blending of conventional and irregular approaches across the full spectrum of conflict . . . future conflict will likely be characterized by a fusion of different forms of warfare rather than a singular approach . . . U.S. forces must become more adaptable and flexible . . . [DoD] officials have discussed the need to counter the continuum of threats that U.S. forces could face from nonstate- and state-sponsored adversaries, including computer network and satellite attacks; portable surface-to-air missiles; improvised explosive devices; information and media manipulation; and chemical, biological, radiological, nuclear, and high yield explosive devices." SOURCE: GAO. 2010. "Hybrid Warfare: Briefing to the Subcommittee on Terrorism, Unconventional Threats and Capabilities," Committee on Armed Services, House of Representatives. GAO-10-1036R. September 10. Available at http://www.gao.gov/new.items/d101036r.pdf. Accessed March 23, 2012.

[19]Ibid.

[20]Andrew Krepinevich. 2011. "The Way to Respond to China." *Los Angeles Times*, November 9. Available at http://articles.latimes.com/2011/nov/09/opinion/la-oe-krepinevich-pacific-20111109. Accessed April 12, 2012.

> The rise of China has triggered a debate among policy experts. On one side sit those who advocate greater engagement. They focus on improving our economic and political relations as the path most likely to maintain stability and peace and . . . those who believe the U.S. and its allies should take steps to offset China's growing military power with the goal of retaining the stable military balance that has benefited all in the region, none more so than China.[21]

Which of these views prevails will still have significant Air Force ISR force-planning implications, especially in the category of doing everything possible to prevent conflict and to prevent a Chinese strategic advantage in the region. Defense policy analyst Andrew Krepinevich has outlined some examples of Chinese aggressive tendencies and provocations, such as the following:

> Chinese fighter jets intercepting and striking a U.S. reconnaissance aircraft in international waters, a Chinese anti-satellite test that created huge quantities of space debris, incidents between Chinese and Japanese aircraft and ships in the East China Sea, and Chinese provocations against Vietnamese oceanographic survey ships in the South China Sea. The objective of China's buildup may not be to wage war. Rather, China may seek to steadily shift the military balance in its favor to the point where Washington can no longer credibly defend either its interests or its allies. In that case, war would not be necessary to ensure China's regional hegemony.[22]

Air Force ISR force planners must consider the concept of future war important, in addition to considerations of how to prevent conflict. In this regard, traditional deterrence models would have to adapt the focus on deterring Chinese superiority within the region short of going to war.

The United States has been in a constant state of readiness on the Korean Peninsula for more than 60 years. The argument can be made, however, that authoritarian dictators can repress their populations for decades to the extent that "confrontational stability" exists.[23] There are at least three potentialities to consider when viewing the Korean Peninsula: (1) a status quo transition from Kim Jong-il to his son Kim Jong-un, (2) an overly aggressive transition that provokes responses from South Korea and the United States, and (3) an accelerated collapsing of the North Korean government. Bruce Bennett and Jennifer Lind argue as follows:

---

[21]Ibid.

[22]Ibid.

[23]General argument made by Bruce Bennett and Jennifer Lind, Fall 2011, *Journal of International Security*, and referenced in "Doomsday War Games: Pentagon's 3 Nightmare Scenarios," *Christian Science Monitor*, December 2011.

> [T]he transition from apparent stability to collapse can be swift. A government collapse in North Korea could unleash a series of catastrophes on the peninsula with potentially far-reaching regional and global effects. This could trigger a massive outflow of the nation's 24 million people, many of whom are severely malnourished, across the border into South Korea . . . Equally troubling, North Korea's weapons of mass destruction could find their way onto the global black market. As a result, the consequences of a poorly planned response to a government collapse in North Korea are potentially calamitous.[24]

The magnitude of a calamitous scenario could quickly outpace a U.S. military response; as noted in the 2010 Quadrennial Defense Review: "The instability or collapse of a WMD [weapons of mass destruction]-armed state is among our most troubling concerns."[25] Such an occurrence could lead to a rapid proliferation of WMD material, weapons, and technology and could quickly become a global crisis posing a direct physical threat. Air Force ISR assets and capabilities would be brought to bear significantly along three operational paths:

1. Maintaining persistent situational awareness of Democratic People's Republic of Korea (DPRK) forces,
2. Detecting precursors to a North Korean missile launch, and
3. Supporting immediate air strikes by both the Republic of Korea and the United States in the event of a North Korean incursion into the South.

The battlefield geography is well known, and North Korea and South Korea share a 238-km border; a conflict along the Demilitarized Zone (DMZ) (roughly 20,000 km^2) would have a combat density greater than that of any engagement in Iraq or Afghanistan. The threat of nuclear engagement is real and must be considered as a worst case in order to deter and, in the event that deterrence fails, plan for full-scale military action. Plans must also be made to manage, in the post-engagement period, the subsequent human tragedy that would unfold for both

---

[24] Ibid.

[25] DoD. 2010. *Quadrennial Defense Review Report.* February. Available at http://www.defense.gov/qdr/images/QDR_as_of_12Feb10_1000.pdf. Accessed April 12, 2012.

North and South Korea as well as for much of the Pacific Rim.[26] The Korean Peninsula is a clear example of Air Force ISR assets being needed at the strategic, operational, and tactical levels, as well as for purposes of pre-war deterrence and the obtaining of early indications and warnings of the potential for a major regional conflict.[27]

Global (Non-Traditional) Scenarios

It appears increasingly clear that the United States will confront a very diverse and demanding array of strategic challenges over the coming decades: transnational terrorist groups, weak and failed states, and the intersection between them; the rise of a near-peer competitor that is not yet overtly hostile toward the United States, but has nonetheless implemented a comprehensive military modernization program devoted to countering the U.S. military's ability to project power; and the proliferation of nuclear weapons to aggressive regimes and perhaps eventually

[26]Arguably, the North Korean regime is essentially stable; it survived its origin in 1950-1953, the collapse of the Soviet Union in the early 1990s, and a devastating famine in 1995-1998. In stark contradiction, authoritarian regimes, like Iran or China, tend to be more unstable. North Korea is a thoroughly totalitarian society, in which all information about the outside world is limited, and dissenting voices are silenced. Although hardship and black markets may undermine the DPRK, there is always China to provide support against disintegration. China has no interest in seeing the DPRK collapse, since doing so (1) may unleash a destabilizing flood of refugees, and (2) its successor state will probably align with, or be absorbed by, South Korea, which is a regional rival and a firm ally of the United States. The Chinese will most likely do everything in their power to avoid a scenario in which a united Korean Peninsula is allied more with the United States than with mainland China. SOURCE: Evan B. Montgomery. 2010. *Defense Planning for the Long Haul: Scenarios, Operational Concepts, and the Future Security Environment.* Center for Strategic and Budgetary Assessments. January 11. Available at http://www.csbaonline.org/publications/2010/01/defense-planning-for-the-long-haul/. Accessed March 23, 2012.

[27]Committee assumptions about Air Force ISR force planning: (1) The primary concern is to detect North Korean missiles being prepped and fueled, so that the United States is capable of intercepting them before launch. (2) Signs of the buildup will likely be detected, and confirmed for certain hours, if not days, in advance. (3) The DPRK's air defense system is extremely dense, and many artillery positions are concealed and/or hardened. (4) The system's obsolescence makes it ineffective against stealth, and it can be easily jammed by modern/existing electronic countermeasures. (5) Although hardened, the ensuing lack of mobility makes them very vulnerable to a full array of precision-guided weapons. (6) Although foliage is still some cause for technical U.S. ISR concern, it is doubtful that concealment will do the North Koreans much good when a majority of hidden artillery positions are identifiable on publicly accessed satellite search engines. Additionally, it must be noted that the threat environment in the Pacific region stands in stark contrast to that in Southwest Asia where currently U.S. remotely piloted aircraft operate without significant restraint from anti-access or enemy denial capabilities. It is also important to note that all U.S. military capabilities would be seriously degraded by a loss of space assets.

to non-state actors and to those nation-states where civil conflict is likely.[28] The cyber domain also presents the challenge of non-conventional attacks. Terrorist cells or organizations could shift their emphasis from killing Americans to injuring Americans financially, with cyberattacks on Wall Street becoming a common occurrence, for example. There are at least three categories of non-traditional scenarios that must be thought through conceptually to ascertain military force-planning capabilities:

1. *Non-military attacks provoking non-military or non-traditional responses:* Examples could be cyberattacks on public or private networked U.S. infrastructure, and other terrorist activity of which the likely origin does not reside in a nation-state and is widely construed as transnational.
2. *Civil conflict in a country in which direct U.S. involvement is problematic*—either because the stakes do not rise to the level of direct harm to the United States, or because the involvement includes the acceptance of significant military risk with the anticipated political outcome being unclear.
3. *Non-combat contingencies requiring U.S. military involvement directly or indirectly:* These could range in magnitude from needs for very localized humanitarian support to massively large-scale responses that would be beyond any single nation-state or region and for which the United Nations humanitarian infrastructure would be ill prepared.[29]

U.S. military capabilities were brought to bear in the recent tragic events in both Japan and Haiti, and in other parts of the world as well. In structuring for humanitarian assistance, AF ISR capabilities of broad scope would be required—capabilities ranging from assisting a foreign government with significant infrastructure resources to assisting a government with degraded infrastructure.[30]

Homeland Security-Based Scenarios

There are numerous hypothetical scenarios involving current DoD military capabilities in support of national and state agencies within current federal law. In such scenarios, the DoD normally plays a supporting role rather than a primary

[28]Bruce Bennett and Jennifer Lind, Fall 2011, *Journal of International Security,* and referenced in "Doomsday War Games: Pentagon's 3 Nightmare Scenarios." *Christian Science Monitor,* December 2011.

[29]George Freidman. 2009. *The Next 100 Years: A Forecast for the 21st Century.* New York: Doubleday.

[30]Japan and Haiti: In March 2011, a massive earthquake triggered the devastating tsunami that hit Japan, causing a tragic chain of events affecting two nuclear power plants at Fukushima; on January 12, 2010, a 7.0-magnitude earthquake tragically devastated Haiti; in both cases U.S. military support was critical in delivering humanitarian aid and assistance.

role.[31] Although there are numerous examples of defense in support of civil authorities' activities, there are at least three major categories in which direct military support is envisioned: (1) physical/territorial/border protection; (2) consequence management of a natural disaster or terrorist-initiated disaster at a significant threshold requiring a homeland federal response; and (3) presidentially directed activity, currently defined by existing law or new proposed legislation.[32] In the homeland security-type scenarios, it will be most important for Air Force ISR force planners to look at adaptive ways to use military assets in a wide-ranging spectrum of activity, but in ways that would always recognize the legal restrictions inherent in the use of such assets. The more robust scenarios would challenge the limits of that civil adaptation so that operational-use challenges are highlighted from a legal viewpoint rather than a technological basis, which is traditional in analyzing a foreign threat.[33] Arguably, any scenario-based viewpoint will assist planners as they both assess the current state of their force planning and capability analysis and develop new analytical techniques and processes. Robust discussion on these potential scenarios is warranted and should naturally undergird all Air Force ISR force planning.[34]

[31]Defense Support of Civil Authorities (or DSCA) is the current process by which United States military assets and personnel can be used to assist in missions normally carried out by civil authorities. These missions have included responses to natural and man-made disasters, law enforcement support, special events, and other domestic activities. DSCA is the overarching guidance with respect to how the United States military can be requested by a federal agency and the procedures that govern the actions of the military during employment. The military can offer a variety of assistance, which includes personnel or equipment. Among the most sought-after assets are transport (land, sea, and air); fuel; communications; commodities including food, building supplies, and medicines; manpower; technical assistance (especially logistics and communications), and the use of military facilities.

[32]The list of laws applicable to DSCA are numerous and complex; for example: Posse Comitatus Act, 18 U.S.C. 1385; Insurrection Act, 10 U.S.C. 331-335; Robert T. Stafford Disaster Relief and Emergency Assistant Act, 42 U.S.C. 5121-5206; Homeland Security Act, 2002, 6 U.S.C. 101; National Emergencies Act, 50 U.S.C. 1601-1651; Economy Act, 31 U.S.C. 1535, have all been used, modified, or are newly created since September 11, 2001, and issues concerning DSCA are consistently being raised.

[33]Two examples may be anticipated: (1) civilian airspace incursion and (2) the legal restrictions on spying on U.S. citizens. Air Force ISR planners would want to participate in some appropriate manner in the development and execution of the Federal Aviation Agency's Next-Generation Air Transportation System, specifically as it applies to management of the National Airspace System. There are inherent challenges to flying aircraft without onboard pilots in both restricted and unrestricted airspace. In the second example, the legal restrictions from using the vast array of ISR capabilities against U.S. citizens must also always be in the forefront of Homeland Security-based ISR.

[34]Giulio Douhet, 1928: "to make a good instrument . . . first have a precise understanding of what the instrument is to be used for . . . and he who intends to build a good instrument of war must first ask . . . what the next war will be like."

ORGANIZATION OF THE REPORT

This chapter provides a broad context of historical factors related to the development of ISR capabilities and considers potential scenarios involving the use of these capabilities. Chapter 2 addresses Task 1 of the TOR by reviewing the current approach to the Air Force corporate planning and programming processes for ISR CP&A. Chapter 3 covers Task 2 of the TOR by reviewing various analytical methods, processes, and models for large-scale, complex domains like ISR, and identifies best practices. Chapter 4 responds to Tasks 3 and 4 of the TOR by offering recommendations for Air Force consideration for the improvement of its ISR CP&A process and an ideal model of an Air Force "system-of-systems" evaluation process for ISR CP&A. Findings are embedded in the text of Chapters 2 and 3 after the supporting evidence.

Appendix A provides biographical sketches of the committee members, and Appendix B presents a list of the meetings held by the study committee, as well as the names of the presenters and participating organizations. Appendix C serves as a supplement to Chapter 3 by providing descriptions of additional organizational CP&A processes and tools.

2

The Current State of the Air Force Intelligence, Surveillance, and Reconnaissance Investment Planning Process

INTRODUCTION

As discussed in Chapter 1, intelligence, surveillance, and reconnaissance (ISR) capabilities enable the U.S. Air Force (USAF) to be aware of developments related to adversaries worldwide and to conduct a wide variety of critical missions, both in peacetime and in conflict. An idealized picture of a global, integrated ISR system is shown in Figure 2-1. It involves a networked system of systems operating in space, cyberspace, air, land, and maritime domains. These systems include planning and direction, collection, processing and exploitation, analysis and production, and dissemination (PCPAD) capabilities linked together by a communications architecture. As suggested in Figure 2-1, different ISR systems may be required in permissive, contested, and highly contested environments.

Although this idealized "enterprise" picture of global, integrated ISR systems is highly desirable, it is not yet treated as an enterprise.[1] ISR systems in different domains tend to be owned and operated by different governmental agencies for the accomplishment of their own particular missions, and even systems operating in the same domain often do not communicate with one another. There is no coordi-

[1]Enterprise" is defined as the set of all U.S. ISR capabilities operating in multiple domains, irrespective of which U.S. agency or organization owns the capability, that are capable of informing decision makers at all levels about the activity of an adversary or potential adversary.

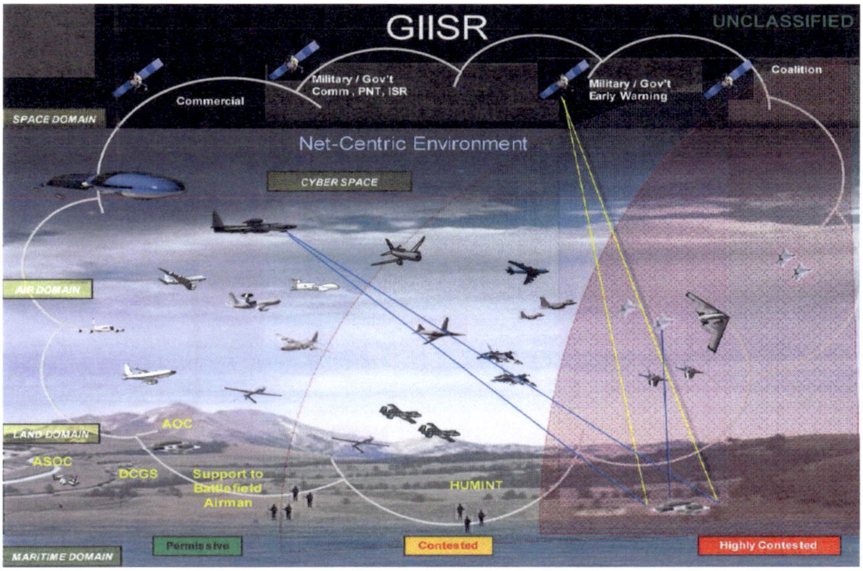

FIGURE 2-1 Global, integrated intelligence, surveillance, and reconnaissance (ISR) operational view. NOTE: Acronyms are defined in the list in the front matter. SOURCE: Col Scot Gere, Chief, GIISR Core Function Team. "Core Function Lead Integrator (CFLI) Construct and GIISR Capability, Planning, and Analysis." Presentation to the committee, January 25, 2012.

nated planning process in place among the many organizations that are stakeholders in ISR systems, and consequently no true enterprise architecture for ISR exists.

This state of affairs is hardly surprising. Generally speaking, the intelligence community (IC) controls the planning and acquisition of national space assets and assets for collecting the various "INTs" (e.g., SIGINT [signals intelligence], HUMINT [human intelligence], among others), while the Air Force and the other military services focus on organizing, training, and equipping forces with ISR capabilities in space, air, and cyberspace (see Box 2-1).[2] Planning and budgeting for ISR missions among these agencies and services are generally done independently; even within a single agency the ISR planning and acquisition programs are often stovepiped, with the resulting systems lacking the standards and common communications systems that would enable them to operate in the coordinated fashion depicted in Figure 2-1.

[2]The IC is composed of 17 member organizations and includes the National Reconnaissance Office, the National Security Agency, the Defense Intelligence Agency, and the National Geospatial-Intelligence Agency. For more information, see http://www.intelligence.gov/about-the-intelligence-community/member-agencies/. Accessed May 24, 2012.

BOX 2-1
The Cyberspace Domain

Cyberspace, a relatively new and rapidly evolving operational domain for the Department of Defense (DoD) and the military services, is defined as "a global domain within the information environment consisting of the interdependent network of information technology infrastructures, including the Internet, telecommunications networks, computer systems, and embedded processors and controllers."[a] ISR can be substantially augmented or hindered in the cyberspace domain. ISR sensors can be augmented by the ability of cyber information to provide geolocation information and movement information on adversarial and friendly systems. This capability can allow sparse assets to be deployed elsewhere or to obtain information more effectively, allowing rapid, minimal observations.

Cyberspace is human-made, which makes the cyber domain different from the natural domains of air and space, although cyber capabilities can exist in all natural domains. Components, subsystems, and systems exist in the cyber domain: these include networks, globally integrated and isolated; physical infrastructure; electronic systems; portions of electromagnetic systems;[b] and industrial control systems known as "SCADA" (supervisory control and data acquisition) systems. The latter are computer systems that monitor and control industrial, infrastructure, or facility-based processes.

Beyond these definitions, the committee offers the view that any asset with computational capability—including avionics and flight control systems, tactical communications and data links, and command-and-control systems onboard and off-board—should be considered to be in the cyber domain.

[a]DoD. 2010. "Department of Defense Dictionary of Military and Associated Terms (Joint Publication 1-02). 8 November. As amended through 15 October 2011." Available at http://www.dtic.mil/doctrine/new_pubs/jp1_02.pdf. Accessed February 6, 2012.

[b]DoD. 2007. "Electronic Warfare." Joint Publication 3-13.1. Available at http://www.fas.org/irp/doddir/dod/jp3-13-1.pdf. Accessed April 12, 2012.

The Deputy Chief of Staff of the Air Force for ISR (AF/A2) and others desire that the Air Force conduct its ISR Capability Planning and Analysis (CP&A) process at an enterprise level rather than on a system-by-system basis.[3] To produce optimum capability, the Air Force wishes to treat ISR data and information as a system-of-systems enterprise. Such an enterprise needs to be composed of end-to-end solutions that include all the elements of PCPAD—planning and direction,

[3]On a system-by-system basis, individual ISR systems are considered in isolation from other ISR systems. The result may be that an ISR system other than the one in question may sufficiently provide the sought-after capability requirement, thus obviating a new acquisition need. Conversely, the system in question may not be needed at all in view of the contribution of another system not considered. Further, the combination of otherwise independently acting systems may together solve the capability requirement. Conversely, in an ISR enterprise, all relevant ISR systems are considered regardless of ownership as long as their capability contributes to understanding an adversary or potential adversary.

FIGURE 2-2 The intelligence process. SOURCE: Department of Defense. 2007. *Joint Intelligence.* Joint Publication 2-0. Washington D.C.: Department of Defense. Available at http://www.fas.org/irp/DoDdir/DoD/jp2_0.pdf.

collection, processing and exploitation, analysis and production, and dissemination—referred to by the Joint Staff as the intelligence process, or the Joint Intelligence Cycle (see Figure 2-2).[4] Deficiencies in any PCPAD element can reduce the effectiveness of the overall intelligence cycle.[5,6] For example, as the Air Force is painfully aware, it does little good to acquire the capability to collect data from a wide-area aerial surveillance system if those data cannot be processed and turned into actionable information within a recognized time period of usefulness.

One of the major issues posed by the integration of new technologies into an existing mix of ISR systems is sustainability. For example, the protracted conflict in Southwest Asia and the demands of the kind of the counterinsurgency (COIN)

[4]DoD. 2007. *Joint Intelligence. Joint Publication 2-0.* Washington, D.C.: Department of Defense. Available at http://www.fas.org/irp/DoDdir/DoD/jp2_0.pdf. Accessed January 4, 2012.

[5]For additional information on the PCPAD process, see Jesse Flanigan, 2011, "Intelligence Supportability Analysis for Decision Making." Available at http://spie.org/documents/Newsroom/Imported/003661/003661_10.pdf. Accessed February 27, 2012.

[6]GAO. 2011. "Intelligence, Surveillance, and Reconnaissance: Actions Are Needed to Increase Integration and Efficiencies of DOD's ISR Enterprise." Available at http://www.gao.gov/products/GAO-11-465. Accessed July 28, 2011.

warfare fought there caused the combatant commander to require a large number of quick-reaction capabilities (QRC) and new ISR capabilities. This quick-reaction, or "Urgent Operational Need" (UON) process worked well in delivering to the warfighter important operational capabilities, but the process is not sustainable in the long run. Because many QRC projects result in the fielding of new technologies and systems for which there is little experience and for which long-term sustainability is an unknown, the costs and difficulties in repairing, training for, supplying, and otherwise supporting a host of one-of-a-kind systems are large challenges for the military services.[7] With the conflict now diminishing, the Air Force needs to determine if it should—or how it should—permanently bring these new capabilities, such as non-traditional ISR (NTISR), into its ISR enterprise.[8,9] The Deputy Chief of Staff of the Air Force for ISR recently noted:

> The Air Force will take "a year or two" to decide whether to keep, expand, or jettison a variety of "boutique" intelligence-surveillance-reconnaissance capabilities created as ad-hoc solutions to special needs during the past 10 years of war in Iraq and Afghanistan. . . . These "quick-reaction capability" programs, such as Gorgon Stare and Blue Devil, to name just two, "need to play out" a while longer so USAF can determine if they are worth the expense of continuing. . . . Gorgon Stare vastly increases the ISR "take" from an MQ-9 Reaper, for instance, but the Air Force is staggering under the weight of the data the systems are generating. . . . Gorgon Stare and Blue Devil generate "53 terabytes a day" of data, equivalent to "12 years of video". . . . Collectively, . . . USAF's high-definition video systems are generating six petabytes, or "80 years" of high-def video a day. USAF will have to invest heavily in processing, exploitation, and distribution systems to keep up with the flow, and will need lots of analysts skilled at synthesizing "all source" ISR. . . .[10]

[7]A report from the National Research Council (NRC) identifies the long-term sustainment of rapid prototypes as a potential major issue. NRC. 2009. *Experimentation and Rapid Prototyping in Support of Counterterrorism.* Washington, D.C.: The National Academies Press.

[8]"Non-traditional ISR" is defined as follows: "NTISR is the concept of employing a sensor not normally used for ISR as part of an integrated collection plan developed at the operational level for preplanned, on-call, ad hoc, and/or opportune collection." SOURCE: USAF. 2007. Air Force NTISR Functional Concept.

[9]The Vice Commander of the Air Combat Command used another example of the need for an ISR Capability Planning and Analysis process that can bend and adjust existing programs of record so that they produce capabilities that take advantage of NTISR. He noted that while new fighter aircraft have immensely powerful ISR collection capability, they lack the ability to get the ISR information into the hands of those who can use it. "*This requires changes in both material and non-material ways in such areas as command and control, data links, processing and dissemination. The Air Force has known this for over a decade, but the ability to describe and adjust to changes to make this NTISR capability a reality has not developed. A faulty CP&A process could be a major factor in why this failure has occurred*" [emphasis added]. SOURCE: Lt Gen William Rew, Vice Commander, Air Combat Command. Personal communication to the committee, January 25, 2012.

[10]John Tirpak. 2012. "Boutique ISR," *Air Force Magazine*, February 16. Available at http://www.airforce-magazine.com/DRArchive/Pages/2012/February%202012/February%2016%202012/BoutiqueISR.aspx. Accessed March 22, 2012.

According to Chairman of the Joint Chiefs of Staff Instruction (CJCSI) 3170.01G, there are three key Department of Defense (DoD) processes—that is, the requirements process, the acquisition process, and the planning, programming, budgeting, and execution (PPBE) process—that need to work in concert to deliver the capabilities required by the warfighter.[11] Each process is ongoing, and keeping all three synchronized has been problematic. The Air Force has long used a variety of methods and tools to evaluate and inform its ISR requirements, acquisition, and PPBE decisions. To ensure that these three key processes are more synchronous, the Air Force has recently undertaken steps to improve its ISR CP&A process. These efforts have gone a long way toward developing more rigor and collaboration in the identification of operational needs and the acquisition of systems.

HISTORICAL DEVELOPMENT OF THE AIR FORCE ISR PLANNING PROCESS

In 2008, the Government Accountability Office (GAO) called for the DoD to have an integrated ISR enterprise architecture and framework for providing and considering trade-offs among future potential investment alternatives.[12] This action has yet to be taken, although the Office of the Under Secretary of Defense (Intelligence) (OUSD[I]) announced in 2010 its intention to create a Defense Intelligence Mission Area Enterprise Architecture. A June 2011 GAO report noted the lack of an implementation plan and time line for this new enterprise architecture.[13] Any Air Force enterprise ISR perspective and architecture would have to be consistent with this Defense Intelligence Mission Area Enterprise Architecture when it is developed. The magnitude of this challenge is depicted in Figure 2-3, which shows the number of organizations that have some responsibility for ISR. The Air Force is but one, albeit large, ISR capability provider to the nation. Decisions made regarding Air Force ISR capabilities need to take into account the organizations listed in Figure 2-3. Many of these key organizations are responsible for creating, evaluating, and using Air Force ISR capabilities.

In providing ISR capabilities, the Air Force is required to make investment decisions that recognize that the requirements for its ISR capabilities come either

[11] CJCSI. 2009. "Joint Capabilities Integration and Development System." CJCSI 3170.01G. Washington, D.C.: Department of Defense. Available at http://www.dtic.mil/cjcs_directives/cdata/unlimit/3170_01.pdf. Accessed February 27, 2012.

[12] GAO. 2008. *Intelligence, Surveillance, and Reconnaissance: DoD Can Better Assess and Integrate ISR Capabilities and Oversee Development of Future ISR Requirements.* Available at http://www.gao.gov/new.items/d08374.pdf. Accessed April 13, 2012.

[13] GAO. 2011. *Intelligence, Surveillance, and Reconnaissance: Actions Are Needed to Increase Integration and Efficiencies of DOD's ISR Enterprise.* Available at http://www.gao.gov/products/GAO-11-465. Accessed July 28, 2011.

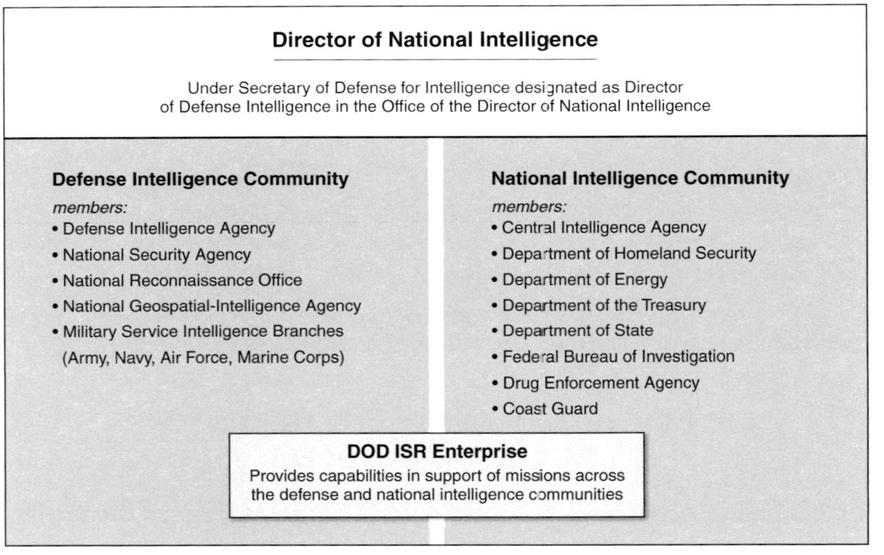

FIGURE 2-3 The relationship of the Department of Defense's intelligence, surveillance, and reconnaissance (ISR) enterprise to the intelligence community. SOURCE: GAO. 2011. Intelligence, Surveillance, And Reconnaissance: Actions Are Needed to Increase Integration and Efficiencies of DOD's ISR Enterprise. Available at http://www.gao.gov/new.items/d11465.pdf.

from Joint combatant commanders or from the nation's top-level decision makers. In the DoD, capability development requirements are vetted through two principal processes, the Joint Urgent Operational Need (JUON)/Urgent Operational Need (UON) process and the Joint Capabilities Integration Development System (JCIDS) process, as they become formalized and validated by senior decision makers in the DoD. The distinction, in principle, is that the JUONs and UONs are intended to be schedule-constrained and limited in scope, size, and potential performance, with a focus on speed to need. The normal process, JCIDS, is intended to ensure rigorous analysis and study in defining the capability need before entering the technology development phase of the acquisition process.

1. JUONs and UONs from the Combatant Commands (COCOMs) or service Major Commands (MAJCOMs) are prioritized by the COCOM/MAJCOM leaders and sent to the Joint Staff and the military services to prioritize and provide solutions according to a "time to field" focus, with a target of less than 12 months to field a solution.
2. In the JCIDS, or the military service-specific capabilities development process for non-Joint requirements, Initial Capabilities Documents (ICDs)

document the required capabilities needed following a Capabilities-Based Assessment (CBA) methodology. In the Joint Staff, the J8 runs the JCIDS process. In the Air Force, the AF/A5 runs the corporate capabilities process through the Air Force Requirements Oversight Council (AFROC), with the Vice Chief of Staff of the Air Force (VCSAF) validating any recommendations. In both JCIDS and Air Force processes, MAJCOMs or COCOMs sponsor the requirements into the processes.

Requirements are formally documented in ICDs and Capabilities Description Documents (CDDs) that must be validated by either the Joint Requirements Oversight Council (JROC), which is chaired by the Vice Chairman of the Joint Chiefs of Staff (VCJCS), or the VCSAF. The JROC publishes its final validation of an ICD in a JROC Memorandum and the AFROC in an AFROC Memorandum. At this point, the requirements are pushed back to the lead MAJCOM for funding and into the Acquisition Systems through the Materiel Enterprise of a particular service (e.g., Air Force Materiel Command) for development.[14] In assessing and planning for its ISR capabilities, the Air Force has to consider the entire set of Joint capabilities provided by the other services as well as the needs of the COCOMs.

The ISR Flight Plan

In 2009 and 2010, AF/A2 produced and used what was called the ISR Flight Plan to articulate how Air Force ISR would meet current and future challenges of air, space, and cyberspace operations and address all doctrine, organization, training, materiel, leadership and education, personnel, facilities, and policy (DOTMLPF-P) considerations.[15] The ISR Flight Plan translated priorities and guidance in the Air Force Strategic Plan to "create a vector for ISR capability development, modernization and recapitalization." It was to be the guiding source for the annual planning and programming guidance (APPG) and was intended, along with the Air Force ISR Strategy, to be the Air Force Core Master Plan for Global Integrated ISR.[16] The major tool used in creating the ISR Flight Plan was the ISR Capabilities Analysis Requirements Tool (ISR-CART), which is a database and searchable repository of requirements and ISR programs and capabilities (see Box 2-2).

---

[14]Lt Col Nathan Cline, ISR Plans and Integration Division, HQ AF/A2DP. Personal communication to the committee, May 24, 2012.

[15]Lt Gen David Deptula (USAF, Ret.), Chief Executive Officer and Managing Director, Mav6. "The Air Force ISR Flight Plan: Origin, Rational and Process." Presentation to the committee, October 6, 2011.

[16]USAF. 2009. "Intelligence, Surveillance, and Reconnaissance (ISR) Flight Plan." Memorandum for ALMAJCOM. June 18. Washington, D.C.: Office of the Chief of Staff.

BOX 2-2
The ISR Capabilities Analysis Requirements Tool (ISR-CART)

The ISR Capabilities Analysis Requirements Tool (ISR-CART) is maintained by the Air Force Intelligence, Surveillance, and Reconnaissance Agency (AFISRA) and sponsored by the AFISRA's A-5/8/9 directorate. The database contains information on many types of intelligence, including mission-related data, from the intelligence community, the other services, the Joint Staff, and industry.

The ISR-CART, an interactive tool, is accessible to authorized users on the Secret Internet Protocol Router Network (SIPRnet) and the Joint Worldwide Intelligence Communications System (JWICS). The SIPRnet is the Department of Defense's (DoD's) classified version of the civilian Internet; it carries information up to and including the Secret classification. JWICS is a similar system of interconnected computer networks primarily used by the DoD, the U.S. Department of State, the U.S. Department of Homeland Security, and the U.S. Department of Justice to transmit classified information at Top Secret or higher levels.

The ISR-CART is the repository of a wealth of information, including nearly all ISR requirements, ISR system attributes and limitations, and ISR capability needs, gaps, and solutions.[a] The database allows users to access information needed to make informed capability and modernization planning decisions and to meet future technology challenges. It provides the ability to link all areas, from stated operational need to proposed solutions, actual research and development (R&D) to delivery of an operational system. ISR-CART has a modularized design enabling links among multiple categories. The modules of ISR-CART include the following: tasks/needs, gaps, solutions, R&D efforts, systems (including parametric information), points of contact, references/bibliography (such as capability guides, concepts of operation), and a glossary.

---

[a]USAF. 2009. "Intel Deputy Unveils ISR Capability Planning Process." Available at http://www.af.mil/news/story.asp?id=123143770. Accessed February 27, 2012.

Guided by both directive and Air Force instruction signed by the Secretary of the Air Force (SECAF) and Chief of Staff of the Air Force (CSAF), the ISR Flight Plan was put together in a collaborative, iterative process led by AF/A2, with representation and subject-matter experts from the MAJCOMs, including the Air Force Materiel Command's (AFMC's) Air Force Research Laboratory, the Air Force ISR Agency (AFISRA), the National Air and Space Intelligence Center, and the Headquarters Air Force (HAF) staff, including AF/A5, AF/A7, AF/A8 and SAF/AQ. The ISR Flight Plan began by considering strategic guidance and then tied in to the results of the ISR work of the capability-based planning (CBP) and the JCIDS processes. The ISR Flight Plan process culminated with a variety of options for HAF consideration in the creation of the Program Objective Memorandum (POM) of the PPBE process.

The ISR Flight Plan took more than a year to complete and required many hours of work from many people in a variety of Air Force organizations. It was well received, as it filled a need in the assessment of ISR capability. Until the ISR Flight Plan, there had been no other attempt at a holistic, across-the-Air-Force examination of ISR requirements, capabilities, needs, gaps, and solutions to yield options to guide ISR planning and programming. However, its completion came just as the Air Force leadership decided to undertake a new method of determining programmatic needs of core Air Force functions.

The Core Function Lead Integrator Construct

In 2010, the Secretary of the Air Force and Chief of Staff of the Air Force decided that each of the Air Force's 12 Service Core Functions would have an annual Core Function Master Plan (CFMP) developed under the guidance of an Air Force MAJCOM commander acting as a Core Function Lead Integrator (CFLI).[17,18] For one of these core functions—Global Integrated ISR, or GIISR—it was determined that the CFLI would be the Air Combat Command (ACC). In 2011, each CFLI was tasked to produce a baseline CFMP. In this work they were to align strategy, operating concepts, and capability development with requirements and programmatic decisions about the Service Core Function over a 20-year period.

The ISR Flight Plan, delivered once, was not updated for the following year, as resources and leadership attention of the Air Force were turned to CFMP production for GIISR and the other CFMPs. AF/A2 staff was left to wonder what to do with the processes of the 2009 ISR Flight Plan and what would be the relationship of the CFLI with the HAF staff responsibilities in capability planning and assessment. Some of the results of the ISR Flight Plan (such as gap analysis)—and some methods and tools (such as the ISR-CART)—were used in the development of the 2010 GIISR CFMP. AF/A2 renamed the ISR Flight Plan process the ISR CP&A process.

---

[17]Service Core Functions define the Air Force's key capabilities and contributions as a service. Service Core Functions correspond to the specific primary functions of the service as described in DoD Directive 5100.01. SOURCE: USAF. 2012. "GIISR Operations. Air Force Doctrine Document 2-0. Dated 6 January." Available at http://www.fas.org/irp/doddir/usaf/afdd2-0.pdf. Accessed March 22, 2012.

[18]Following are the names of the Air Force Service Core Functions: (1) Nuclear Deterrence Operations, (2) Air Superiority, (3) Global Precision Attack, (4) Personnel Recovery, (5) Command and Control, (6) Global Integrated ISR, (7) Space Superiority, (8) Cyberspace Superiority, (9) Rapid Global Mobility, (10) Special Operations, (11) Agile Combat Support, and (12) Building Partnerships. SOURCE: Col Brian Johnson, Chief, Air Force Plans and Integration Division for the Deputy Chief of Staff, Intelligence, Surveillance and Reconnaissance, Headquarters, U.S. Air Force. "Air Force ISR CP&A Overview." Presentation to the committee, October 6, 2011.

In summary, the Air Force employs two overlapping processes for planning future ISR investments: the ISR CP&A process, which is derived from the earlier ISR Flight Plan and led by the AF/A2; and the CFLI process, led by the Air Combat Command. These processes are described below. At this writing, the two processes have not been fully reconciled, with consequences that are discussed later in this chapter.

THE CURRENT AIR FORCE ISR CAPABILITY PLANNING AND ANALYSIS PROCESS

The current Air Force ISR CP&A process, as shown in Figure 2-4, is informed and guided by strategic direction provided by the White House National Security Council, the U.S. Congress, the DoD, the IC through the Director of National Intelligence (DNI), and others. This guidance is handed down at different times and takes many written forms, including National Intelligence Estimates, the Five-Year Defense Plan, Global Threat Analyses, and various global trend studies often conducted by Federally Funded Research and Development Centers.

The Needs Analysis phase of the CP&A process attempts to ensure that *all* ISR needs are gathered from across the Air Force ISR enterprise. Primary participants in the Needs Analysis phase are AF/A2 and its direct reporting organization, AFISRA, as well as the COCOMs, which typically express their needs through Integrated Priority Lists, and the MAJCOMs, which represent the interests of their affiliated COCOMs. The National Reconnaissance Office (NRO) and its mission partners, the National Geospatial-Intelligence Agency (NGA) and the National Security Agency (NSA), which have their own capability planning and analysis processes, appear to engage only tangentially in the Air Force process.

The Needs Analysis function produces an unconstrained, "1-to-N" list of ISR needs. No attempt is made at this step in the current process to prioritize or filter needs. By gathering *all* needs, the process seeks to prevent needs that might not be relevant in today's mission environment from falling on the cutting-room floor. However, gathering all needs each time through the process can be very time- and labor-intensive and may not be necessary for situations in which investment decision makers are interested in rapid answers to focused questions.

The Gap Analysis function matches each need on the list with known ISR capabilities. Needs that have no matching capabilities are identified as gaps. Major participants in this phase of the process include AF/A2, AFISRA, the GIISR CFLI, and MAJCOM representatives. As with the Needs Analysis phase, participation by the IC appears to be more opportunistic than systematic.

The primary tool used to match capabilities with needs is the ISR-CART database. (See Box 2-2.) ISR-CART, which is maintained by AFISRA, provides a comprehensive, searchable store of needs, capabilities, and gap information, indexed by a variety of metadata types. Although the physical process of matching needs

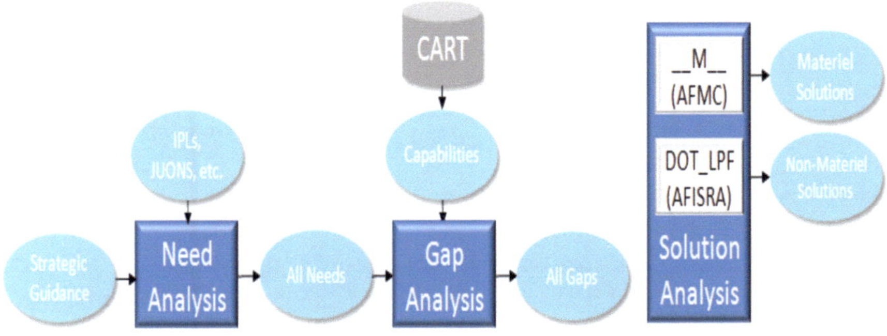

FIGURE 2-4 Major elements, inputs, and outputs of the current Intelligence, Surveillance, and Reconnaissance Capability Planning and Analysis (ISR CP&A) process of the Deputy Chief of Staff of the Air Force for ISR (AF/A2). NOTE: Acronyms are defined in the list in the front matter. SOURCE: Derived from Col Brian Johnson, Chief, ISR Plans and Integration Division (AF/A2DP), Headquarters, U.S. Air Force, "Air Force ISR CP&A Overview." Presentation to the committee, October 6, 2011.

with capabilities is completely manual, and therefore labor- and time-intensive, the ISR-CART information repository serves as a highly valuable "one-stop shop" for obtaining vital information needed to support the Needs and Gap Analysis processes.

The first, and to date apparently only, pass through the Gap Analysis phase of the ISR CP&A process yielded a well-documented list of more than 200 gaps. Items on this list were further grouped into 12 prioritized categories.[19] The Solutions Phase then began the process of analyzing potential solutions for each gap category. Materiel solutions were investigated by AFMC, and gap areas were assigned to appropriate Capability Management Teams that included representatives from various Air Force science and technology and acquisition stakeholder communities.[20] Non-materiel solutions, or so-called DOT_LFP solutions, were investigated under the leadership of AFISRA.

Although AFMC and AFISRA made valiant attempts to investigate materiel and non-materiel solutions, the process appears to have become bogged down by the lack of manpower and funding resources required to adequately investigate more than a small number of gap areas. In addition, the ISR CP&A process was paused at the request of the ACC as the GIISR CFLI stood up and began the process

[19]Col Brian Johnson, Chief, ISR Plans and Integration Division (AF/A2DP), Headquarters, U.S. Air Force. "Air Force ISR CP&A Overview." Presentation to the committee, October 6, 2011.

[20]Brig Gen Dwyer Dennis, Director, Intelligence and Requirements Directorate, Headquarters AFMC, Wright-Patterson Air Force Base, Ohio. "Capture the Past, Build the Future: Capability Planning and Analysis." Presentation to the committee, November 10, 2011.

of defining the roles and responsibilities.[21] Following this summary of the current process is a more in-depth look at both how this process came to be and what it is.

THE CORE FUNCTION LEAD INTEGRATOR PROCESS

As noted above, in 2010 the Air Force undertook a new service-wide capabilities-based planning method as part of a revised strategic planning process. At the heart of this process are annually iterated CFMPs developed by Air Force MAJCOM commanders who act as CFLIs for specific Service Core Functions. The CFLI is the authoritative source for detailed planning within each Service Core Function. As stated above, the CFLI for the GIISR core function is the commander of the Air Combat Command.

There are three aspects of the Service Core Function construct that are noteworthy. First, CFMPs for two Service Core Functions—GIISR and Command and Control—are unique among CFMPs because they are enablers for all other Service Core Functions. Second, the Space Superiority and Cyber Superiority Service Core Functions, both of which have strong connections to the ISR enterprise, are led by a different CFLI (Air Force Space Command [AFSPC]) than the CFLI that leads the Service Core Functions for GIISR and Command and Control. Third, the management of those items that would constitute NTISR is fragmented among other ACC CFMPs. The potential exists for ISR capabilities to be undervalued, underfunded, or completely missed by a given CFMP. If the Air Force wishes to integrate NTISR collection capability from platforms, such as its newest fighters, the CFMP or ISR CP&A processes may have to point to the budgetary choices among several non-ISR programs in order to pay for such capability. For example, the Air Force will have to ensure that the platforms have necessary data links and that the command-and-control structure is capable of tasking the platforms in both near real time and real time, and a capability will be needed to turn the data collected into actionable information in order to support PCPAD. The elements of this example are each in separate CFMPs. It seems that none of the CFMPs has the priority to make NTISR a reality; none has lead responsibility in this fragmented structure.

As the basis for its work in producing the 2011 GIISR CFMP, the Air Combat Command started with the Gap Analysis and the 19 ISR Gap Focus Areas that had resulted from the 2009 ISR Flight Plan. Its next step was to conduct an assessment of risks involving these gaps as applied to three representative scenarios found in operational plans, each of which require Air Force GIISR support. This analysis was constrained both by external guidance and by the number and type of ISR capabili-

[21]Col Scot Gere, GIISR CFT Chief, Air Combat Command, Langley Air Force Base. "Core Function Lead Integrator (CFLI) Construct and GIISR Capability, Planning, and Analysis." Presentation to the committee, January 25, 2012.

ties considered likely to be available for each scenario examined. This analysis was followed by a determination of trade-space priorities done iteratively with AFMC and other stakeholders to determine the types of forces needed and how best to sustain, replace, and improve these capabilities, along with the associated costs. This yielded a list of prioritized capability gaps and science and technology efforts that was itself refined and adjusted by a council of knowledgeable colonels from across the Air Force. The work was then passed between the council of colonels and a solutions working group for pre-acquisition capability planning and analysis or on to developmental planning to produce relevant materiel and/or non-materiel solutions.[22]

Figure 2-5 shows the GIISR CFLI's view of the ISR process for developing planning, programming, and requirements outputs and depicts the relationships of various major processes and the products that flow from or into these processes. The ring of activities in the middle of the chart shows the relationship of Gap Analysis, non-Air Force POM analysis, GIISR CFMP development, and the solutions vector. In Gap Analysis, ISR gaps are collected and reviewed by all ISR stakeholders and consolidated into the ISR-CART database maintained by AFISRA. In the non-Air Force POM analysis, AF/A2 acts as the service interface with the IC and others outside the Air Force and has the preferred vantage point for understanding gaps in and outside the Air Force. AF/A2 is also required to influence and interpret the Quadrennial Defense Review and the Defense Planning Guidance and to produce its own guidance as the ISR Capability Portfolio Manager.

Informed by external guidance and the current annual planning and programming guidance (APPG), the GIISR CFLI applies a scenario-based assessment to link gaps to operational and force management risk and then prioritizes areas for solution work. The CFMP also creates GIISR planning force proposals that facilitate Air Force integration for a balanced POM submission. Courses of action developed from previous solution work are inserted into programmatic action while at the same time requirements are developed and updated. In the solutions vector, the council of colonels from stakeholder organizations reviews the prioritized areas from the CFMP, considers the national inputs from AF/A2, and provides a vector for capability working groups. These groups collect possible materiel and non-materiel solutions and present their findings in the form of courses of action. The working group may also recommend JCIDS actions to drive developmental planning requests that can be undertaken. Note the outer concentric rings in Figure 2-5

[22] A "council of colonels" is a term not officially defined; however, it is understood to mean a council of persons of that rank who represent the interests and perspectives of their various organizations in a discussion or a decision-making forum about what that group believes about a certain issue or matter. Their views are then forwarded to those in higher authority for either information or further deliberation.

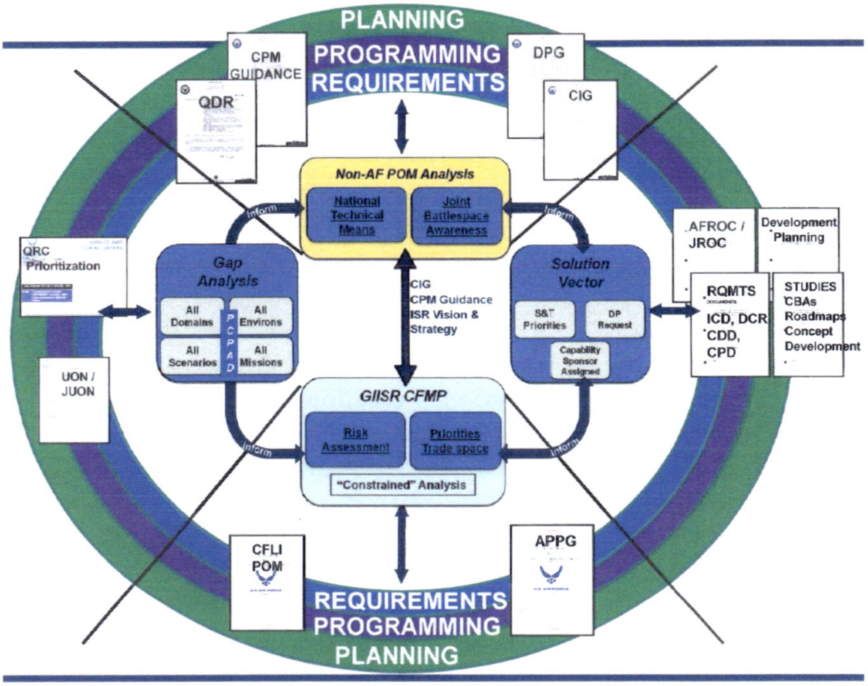

FIGURE 2-5 Intelligence, surveillance, and reconnaissance (ISR) planning, programming, and requirements outputs. NOTE: Acronyms are defined in the list in the front matter. SOURCE: Col Scot Gere, Chief, GIISR Core Function Team. "Core Function Lead Integrator (CFLI) Construct and GIISR Capability, Planning, and Analysis." Presentation to the committee, January 25, 2012.

indicating the relationship of these processes with planning, programming, and requirements processes.

Both the CFMP and the ISR CP&A processes have positive attributes as well as areas for improvement. On the positive side, they are generally inclusive, make a strong effort to use data to inform discussions, and do the best that one could do with an approach that is nearly all manual and labor-intensive. However, the CFMP process, like the ISR Flight Plan before it, is cumbersome and slow and cannot rapidly respond to changes in guidance, urgent warfighter needs, or commanders' needs for quick answers to specific questions. As with the ISR Flight Plan, the GIISR CFLI work took months, consumed many hours of work by subject-matter experts, and utilized no tools other than parametric analysis and the ISR-CART. Further, it was necessary to cross-check with the CFLI staffs developing the CFMPs for Space Superiority and Cyberspace Superiority to eliminate underlap and overlap with the ISR needs of those core functions, and it is not clear to the committee whether or not the underlap/overlap analysis was adequate and correct given the time constraints of the CLFI process.

Space Superiority and Cyberspace Superiority Core Function Master Plans

The Air Force Space Command is the CFLI for the Space Superiority and Cyberspace Superiority Service Core Functions, both of which have ISR content.

Space Superiority

In conducting its work as the Space Superiority and Cyberspace Superiority CFLI, the AFSPC conducts its own CP&A process, which parallels the ISR CP&A process. The AFSPC did participate in the ISR CP&A needs and gap analysis process in 2010. However, there existed some confusion over roles—for example, budget authority—and whether AF/A2 had the clear enterprise role to lead the definitive plan for the Air Force ISR portfolio. The AFPSC believed that AF/A2 had budget authority in the ISR CP&A planning process in the first round of the CFLI process, showing confusion over roles and responsibilities. It seems that more clarification and communication are required among those CFLIs whose responsibilities overlap in ISR capabilities, specifically Space Superiority, Cyberspace Superiority, and GIISR.

Cyberspace Superiority

Owing to the emerging nature of cyberspace operations, the committee offers additional analysis on the concept of Air Force cyberspace operations, the role of the Air Force in the context of the overall DoD/IC cyberspace enterprise, and the relationship of the Air Force to the ISR CP&A and CFLI planning processes. There is a multidimensional relationship between the ISR and cyber missions and capabilities. There are three missions from a cyberspace perspective: support, defense, and force application. ISR is a crosscutting capability that can be applied holistically with other core functions to enable cyberspace missions. Conversely, Cyberspace Superiority supports and is supported by all of the other Air Force core functions. In the case of the GIISR core function, these relationships could be characterized as "Cyber for ISR" and "ISR from Cyber."

The "Cyber for ISR" relationship is illustrated by the mission assurance requirement for the cyber domain in support of an ISR mission. Cyberspace mission assurance ensures the availability and defense of a secured network to support a military operation. If the military operation is an ISR mission, the PCPAD component is reliant on a secured cyberspace infrastructure for communication and dissemination. This dependency should define requirements from the GIISR core function to the Cyberspace Superiority core function. For example, the Air Force currently uses commercial communications segments in some portion of nearly all missions. Short of reconfiguring all communications infrastructure to government-off-the-shelf (GOTS) technology for Air Force missions, this would suggest that requirements for mission resiliency across a hybrid commercial-off-the-shelf (COTS)/GOTS cyber infrastructure in support of ISR missions should be included in a "Cyber for ISR" portfolio planning process—that is, a GIISR CFLI to Cyberspace Superiority CFLI interaction.

Conversely, the "ISR from Cyber" relationship is illustrated by considering how ISR can be executed during cyberspace operations, particularly during cyberspace force application (exploitation). This can be characterized as situational awareness during and in support of cyberspace operations. AFISRA provides all-source cyber-focused ISR including digital network analysis to the 24th Air Force through the 659th ISR Group to enable 24th Air Force operations.[23] AFISR's support includes the following:

1. Current intelligence and reporting,
2. Indications and warning,
3. Threat attribution and characterization,
4. Intelligence preparation of the operational environment, and
5. Computer network exploitation.

Cyberspace ISR requirements are addressed by the Cyberspace Superiority CFLI that, in turn, generates the Cyberspace CFMP. This is another case of ISR requirements being spread across multiple CFLIs. These same ISR requirements could be included in the GIISR CFLI. Cyberspace ISR portfolio planning is part of the 24th Air Force/A2 mission. Although much progress has been made in a relatively short period of time, the 24th Air Force/A2 is still lacking an institutionalized approach to planning and equipping. It is also clear that the required response time for cyberspace ISR capabilities needs to be more rapid than the standard 2-year planning cycle. Moreover, standards and key performance parameters have yet to be identified.[24]

[23]USAF. 2010. "Cyberspace Operations. Air Force Doctrine Document 3-12." Available at http://www.fas.org/irp/DoDdir/usaf/afdd3-12.pdf. Accessed February 27, 2012.

[24]Col Tom French, Chief of ISR Strategy, Plans and Operations (A2X/O), Headquarters AFSPC. "Evolving Cyberspace ISR Corporate Planning." Presentation to the committee, December 8, 2011.

Integration of Air Force Core Function Master Plans

From August to October annually, the Air Staff conducts the integration of all 12 CFMPs. In so doing, it produces cross-service core function/portfolio trades as recommendations on current and future capability needs and investments. CFMP integration (also) identifies Program Force Extended (PFE) program candidates for support or adjustment, as it merges individual CFMP planning force proposals into a unified, fiscally constrained planning force that establishes a 20-year major investment plan. The planning force is then published in the APPG.

ISR Capability Planning and Analysis and Core Function Master Plan Link with Planning, Programming, Budgeting, and Execution

The next consideration in this discussion is how the ISR CP&A and the CFMP link with PPBE. This should begin with a discussion of how the Air Force organizes itself to produce a balanced, annual input to the DoD budget, which is submitted to the Congress for review, adjustment, approval, and funding. Although the SECAF and CSAF make final decisions about the annual POM submission, they rely on something called the Air Force Corporate Structure (AFCS) to do the work of balancing competing demands and managing resource limitations to produce the right PPBE decisions.

The AFCS has several echelons. At the top is the Air Force Council, which is chaired by the Vice CSAF and consists of three-star Deputy Chiefs of Staff. Below that is the Air Force Board of two-star Air Staff generals, and finally the Air Force Group of one-star generals and colonels. These flag officer bodies are themselves supported in issue formulation by a number of mission and mission-support panels.

The flow of POM issues to the CFLI process and the AFCS processes begins with calls for issues by the MAJCOMs and AFISRA (see Figure 2-6). Such issues and requirements are received from throughout the AF and HAF and are collected and represented by Capability Advocates, who review the issues and ensure that they are ready to be brought into decision processes of the MAJCOM or AFISRA. The issues are reviewed and validated with a recommended course of action and then prioritized. The panel's recommendations are then reviewed and validated or modified and then approved by the MAJCOM or AFISRA. The list of approved issues is parsed into investment issues (destined for the CFLI process) and organization and management (O&M) issues (destined for the HAF process).

When the MAJCOM process is complete, the investment issues are forwarded to the appropriate CFLI. The CFLI then takes briefings from knowledgeable staff officers—program element monitors who keep daily track of issues and of available funding for their programs. Issues, with a prioritization and developed course of action, then are to be reviewed by the MAJCOM or AFISRA. Issues and offsets are

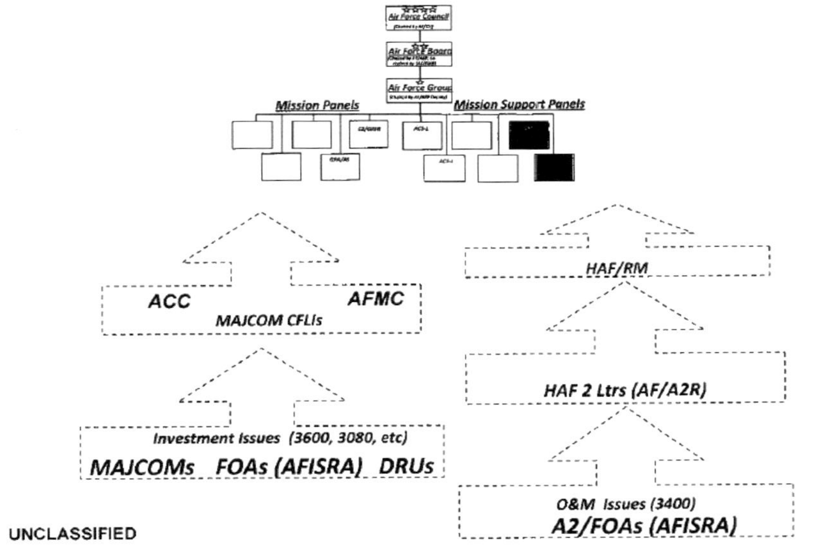

FIGURE 2-6 Program Objective Memorandum (POM) issue routing to the Air Force Corporate Structure. NOTE: Acronyms are defined in the list in the front matter. SOURCE: U.S. Air Force.

both considered, as the CFLI has to present a balanced submission to the Air Staff's AFCS panel responsible for that particular portfolio. The issues are then approved by the CFLI. Once the list is approved, it goes directly to the appropriate Air Staff panel for the beginning of deliberations in the AFCS. Issues are prioritized by AF/A2 and submitted to Headquarters Air Force Resource Management (HAF/RM). HAF/RM presents a resource balanced portfolio to the Air Staff leadership. It does not necessarily present a portfolio having optimized ISR capabilities. This lack of optimization is generally the result of the need for the Air Force to take from one element, such as an ISR need, in order to pay for a more pressing non-ISR need.

Linkages Between the Air Force and the Intelligence Community

AF/A2 is the focal point for Air Force interaction with the IC. Although the DoD has a rigorous and well-defined process for requirements development, the IC is not monolithic, and its process is by necessity considerably less procedural than that of the DoD. It should also be remembered that although the capabilities developed by the two communities may be similar, their uses and the funding used to procure them are different. Specifically, DoD intelligence systems are funded through the Military Intelligence Program (MIP), whereas national intelligence systems are funded through the National Intelligence Program (NIP). In some cases, MIP funding is transferred to individual intelligence agencies to acquire specific capabilities. AF/A2 interactions are, therefore, often point-to-point with individual intelligence agencies. Separate offices within intelligence agencies will work with military users, but each agency has a centralized office with responsibility for supporting combatant commanders and military users. The NRO, for instance, has a Directorate for Mission Support that coordinates support and provides deployable teams to various military commands. AF/A2 coordinates individually to determine where there may be synergy and overlap in Air Force and agency investments in ISR capabilities or where data sharing and collaboration may mitigate further service or agency investments.

The Air Force plays an important role in the IC that lies under the purview of the Director of National Intelligence (DNI). The role of the Air Force in the IC includes the following: (1) weapons systems analysis, particularly air- and air-defense-related all-source analysis, provided by the National Air and Space Intelligence Center; (2) participation in the NSA's cryptologic activities as a Service Cryptologic Element, accomplished as part of the mission of the Air Force ISR Agency; (3) the acquisition and operation of a variety of national-level ISR capabilities, including Cobra Judy and Cobra Dane, as part of the DNI's General Defense Intelligence Program; and (4) the articulation by AF/A2 of the value and importance of IC collection against particular potential threats that has led to IC acquisition of the new and successful systems. Moreover, a variety of Air Force reconnaissance platforms, such as U2s, Rivet Joint aircraft, and overhead persistent infrared space-based capabilities, are regularly used to address DNI requirements. Together, these activities involve a significant amount of the Air Force budget and thousands of Air Force personnel, both military and civilian.

Finally, as noted earlier, AF/A2 is the primary interface between the Air Force and the Office of the Director of National Intelligence for planning and funding those ISR capabilities that are part of the NIP. However, NIP-funded programs have been excluded from Air Force Total Obligation Authority (TOA) review. Thus, the Air Force needs increased awareness of what capabilities it provides, along with the IC and other services, to the Joint fight to reduce duplication of effort and funds expended. Given the large amount of resources included in Air Force TOA for national intelligence activities, there should be considerably more attention given to this issue in the Corporate Air Force process beyond AF/A2.

FINDINGS

Developing an enterprise approach to ISR investment planning is a difficult and complex challenge, and the Air Force processes that have been put in place to wrestle with this challenge are relatively new and still evolving. As expected with new processes aimed at complex problems, there are deficiencies that need to be addressed.

Finding 2-1. The responsibility for evaluating and informing decisions about Air Force ISR capabilities is diffuse, overly personnel-intensive, and divided among many organizations, resulting in an excessively lengthy process. Specifically, the respective roles and responsibilities of the AF/A2 and the GIISR CFLI are not well defined or well understood, and appear disconnected. Both the ISR CP&A and the CFLI processes have positive aspects, but the processes are immature and insufficiently integrated.

It appears that there are conflicting views held by AF/A2 and the GIISR CFLI regarding roles and responsibilities. The Air Combat Command stressed that the CFLI is charged with producing an annual CFMP and that AF/A2 should only provide Gap Analysis into the CFLI process, which then carries out Solution Analysis. However, the absence of guidance about the relationship between these two organizations has created counterproductive uncertainty. Further, this is exacerbated by frequent changes in process, roles and responsibility, and key personnel.

Finding 2-2. The Air Force ISR planning process lacks adequate process definition and formal interaction between the Space Superiority, Cyberspace Superiority, and GIISR CFLIs. It also does not rigorously integrate ISR contributions from other military services, the IC, and the Office of the Secretary of Defense. Consequently, the Air Force process does not yield ISR investment priorities across domains and security constructs. The Air Force needs increased awareness of what capabilities it provides, along with the IC and other services, to the Joint fight to reduce duplication of effort and funds expended.

Finding 2-3. Air Force platforms do not appear to be included in Air Force cyberspace-related planning processes, even though cyberspace vulnerabilities do exist onboard platforms and in the connectivity between them. Moreover, cyberspace functions can play a very positive role in support of ISR, and ISR systems can help support cyberspace functions. Additionally, the complexity of the multi-organizational relationships involved in current DoD and IC interactions leads to confusion in both execution and planning processes, particularly for cyber operations.

Finding 2-4. The Air Force lacks integrated modeling and simulation and analysis tools that provide traceability from requirements to capability and that conduct operationally relevant ISR trade-space analysis across the DOTMLPF-P framework and within and across air, space, and cyberspace domains.

The committee heard about parametric analysis carried out by subject-matter experts. However, the Air Force lacks integrated tools that (1) collaboratively capture operational shortfalls; (2) prioritize needs; (3) realistically portray existing capabilities; (4) identify funding requirements and potential investment trade-space areas; (5) provide the ability to conduct CFLI-focused and ISR corporate-level "what if/if then" drills to assess operational impact and critical-path CFLI investment areas and flow; and (6) provide the ability to determine and recommend the most suitable course of action to maximize ISR capability (across DOTMLPF-P) for and across each Air Force warfighting domain.

Finding 2-5. The Air Force corporate process "disassembles" the ISR portfolio planning analysis, classifies the elements into isolated, or stovepipe, function components, and then makes trade-offs and/or decisions without the ISR trade-space underpinnings.

Finding 2-6. The ISR CP&A process lacks the ability to respond in a timely way with appropriate fidelity to meet the increasing speed of technology development, operational requirements, and the required decrease in planning-cycle time, particularly in the cyberspace domain.

Finding 2-7. PCPAD is not adequately considered and prioritized by the ISR CP&A process.

Finding 2-8. The ISR CP&A process does not adequately consider affordability in capability trade-space analysis.

Table 2-1 summarizes a set of shortfalls that the committee identified in the Air Force ISR CP&A process, aligned with the findings presented in Chapter 2. The Air Force has made great strides in developing the earlier ISR planning processes (ISR Flight Plan, ISR CP&A, and CFLI/CFMP processes). It is the committee's view, however, that improvements can be made to achieve greater efficiency in resource utilization, greater effectiveness in the quality of capability solution determination, and more responsiveness in terms of timeliness and in delivering tailored analysis for the mission solution sought.

CONCLUDING THOUGHTS

Over the past few years, the Air Force has made a significant, concerted effort to organize a comprehensive planning process for the Air Force ISR portfolio. The evolution of this planning process began with the ISR Flight Plan, which was rapidly overtaken by the CFLI/CFMP process. The current processes strive to be very inclusive and collaborative, utilizing cross-ISR community subject-matter experts. These processes also include some coordination across relevant CFLIs and

TABLE 2-1 Air Force ISR Capability Planning and Analysis (CP&A) Process Shortfalls and Corresponding Findings

ISR CP&A Process Shortfalls	Finding
Current process does not adequately address all ISR missions, domains of air, space, and cyberspace managed by the Space Superiority, Cyberspace Superiority, and Global Integrated ISR CFLIs, as well as contributions from other military services, the IC, and OSD, and NTISR capabilities.	Findings 2-2 and 2-3
Current process does not provide the ability to analyze investment decisions at different resolutions and timescales.	Finding 2-6
Current process does not support "what if" analyses in well-defined trade spaces.	Findings 2-4 and 2-5
Current process is too air platform-centric and has insufficient focus on PCPAD.	Finding 2-7
Current process does not adequately address affordability, including acquisition and life cycle, as part of capability trade-space analysis.	Finding 2-8
Current process does not provide traceability from requirements to capabilities.	Finding 2-4
Current process is manual and very labor-intensive, resulting in inefficient use of limited resources.	Finding 2-8
Current process is vulnerable to the inevitable changes in Air Force leadership, organization, strategy, and budgets.	Finding 2-1

NOTE: Acronyms are defined in the list in the front matter.

associated MAJCOMs. And, as a side product of this effort, a very comprehensive repository of ISR needs, capabilities, and gaps has been developed and is now stored in ISR-CART. Still, a number of improvements can be made to the process itself, the analytical tools, models and simulations that can be applied to the process, the emphasis and inclusion of capabilities from across all domains and architectural elements in the process, and the inclusion of other key decision parameters such as affordability. Such improvements would result in a more efficient and effective process and higher-quality outcomes.

The following chapters examine (1) the corresponding processes used in the other services, the IC, and private-sector organizations, with a view to identifying best practices that could be applied to improve the Air Force ISR CP&A process (Chapter 3); and (2) recommendations for improvements to the Air Force process and a proposed future planning process that addresses the shortfalls identified above (Chapter 4).

3

Examples of Processes Employed by Government and Industry for Providing Capability Planning and Analysis

INTRODUCTION

This chapter addresses part 2 of the terms of reference for this study: "Review various analytical methods, processes and models for large scale, complex domains like ISR [intelligence, surveillance, and reconnaissance] and identify best practices." The chapter also discusses government and industry capability planning and analysis (CP&A)-like processes and associated tools, with the aim of identifying attributes and best practices that might be applied to the Air Force ISR CP&A process. The following sections provide brief descriptions of the processes and tools used by several organizations to showcase salient attributes and illustrate best practices of each. Appendix C contains descriptions of additional organizational processes and tools that do not appear in this chapter.[1] At the end of the chapter, Table 3-4 correlates the findings in this chapter with best practices.

EXAMPLES OF GOVERNMENT PROCESSES FOR PROVIDING CAPABILITY PLANNING AND ANALYSIS

The scope of Air Force responsibilities to provide global, integrated ISR capabilities across strategic, operational, and tactical missions is extraordinarily broad

[1]The descriptions of the individual organizations' CP&A-like processes and tools vary considerably and are the result of the intent to provide an unclassified report. Much of the information provided to the committee during its data gathering was classified or otherwise not releasable to the public.

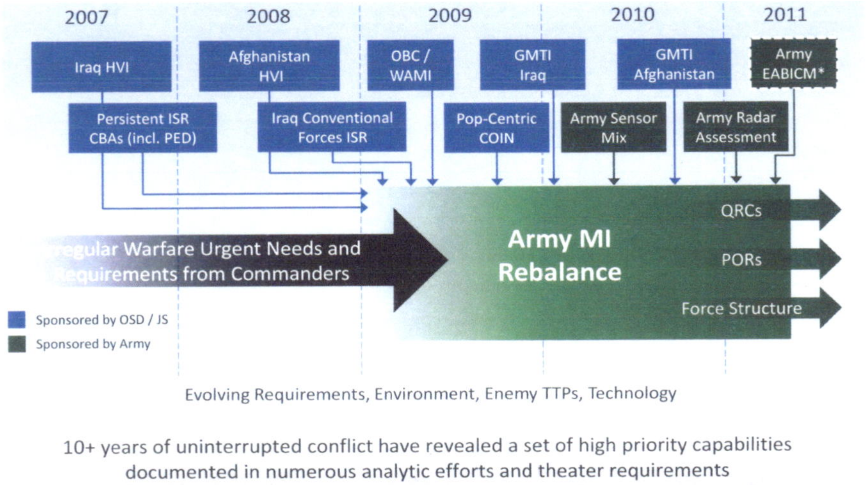

FIGURE 3-1 Analytic underpinnings of intelligence, surveillance, and reconnaissance (ISR) force sizing for the Army. NOTE: Acronyms are defined in the list in the front matter. SOURCE: LTG Richard Zahner, Deputy Chief of Staff, G-2, Headquarters, U.S. Army. "Military Intelligence Rebalance." Presentation to the committee, November 9, 2011.

and complex. Although the organizational processes described below apply, for the most part, to arenas with smaller scope and less complexity, each process was reviewed with the goal of identifying best practices and tools that the Air Force might consider incorporating into its own CP&A process.

U.S. Army

The U.S. Army developed a strategy to rebalance the Army Military Intelligence (MI) Force after a decade of intense ISR system development and deployment in support of operations in Iraq and Afghanistan.[2] This protracted period at war resulted in many system deployments accomplished with great urgency as Quick Reaction Capabilities (QRC), depicted in Figure 3-1. The overarching strategy for Army Intelligence is to optimize core intelligence capabilities in support of Brigade Combat Teams (BCTs) and division and corps full-spectrum operations on a sustained Army Force Generation (ARFORGEN) cycle.[3] Thus, the Army's approach relies principally on its own organic ISR capability rather than on Air Force or na-

[2]U.S. Army. *A Strategy to Rebalance the Army MI Force—Major Themes and Concepts*. Available at http://www.dami.army.pentagon.mil/site/G-2%20Vision/nDocs.aspx. Accessed February 29, 2012.

[3]LTG Richard Zahner, Deputy Chief of Staff, G-2, Headquarters, U.S. Army. "Military Intelligence Rebalance." Presentation to the committee, November 9, 2011.

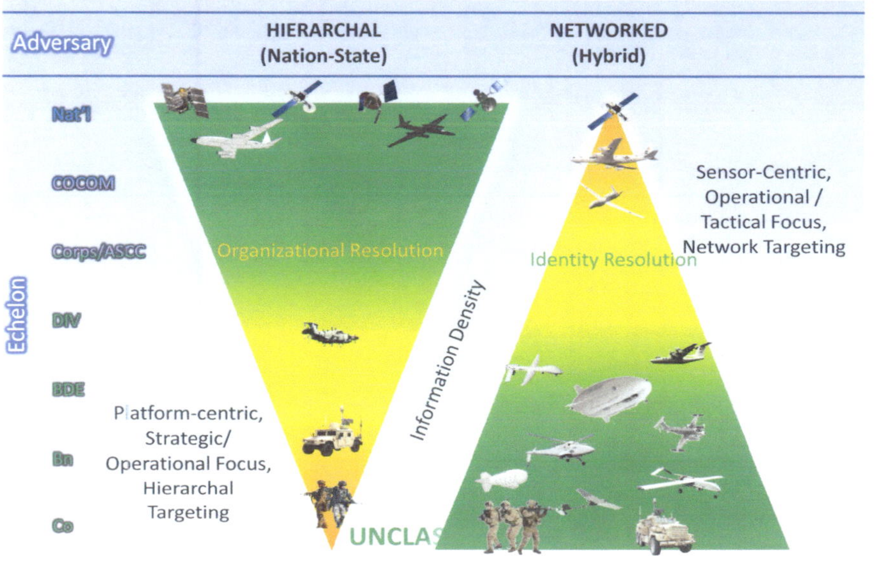

FIGURE 3-2 The Army's intelligence, surveillance, and reconnaissance (ISR) requirements and information density generation in past and in present and future threat environments. SOURCE: LTG Richard Zahner, Deputy Chief of Staff, G-2, Headquarters, U.S. Army. "Military Intelligence Rebalance." Presentation to the committee, November 9, 2011.

tional capabilities. Looking to the future, the MI rebalance is intended to determine which capabilities are enduring, using a Doctrine, Organization, Training, Materiel, Leadership and Education, Personnel, and Facilities (DOTMLPF) assessment.[4]

Figure 3-2 identifies the Army's ISR requirements and information density generation for past threat environments compared with those for present and future threat environments. Cold War requirements were hierarchical and focused on the operational level, whereas contemporary requirements are networked, with a tactical focus. Additionally, for the most part, the Army has recently faced a benign air threat, as coalition forces enjoyed air superiority in Iraq and Afghanistan. This led the Army to focus ISR support more toward tactical units, which are at present and can be expected in the future to prosecute much of the fight.

In support of these decentralized and networked operations, Army Intelligence devised the Integrated Sensor Coverage Area (ISCA) construct, featuring three distinct ISR mission sets, shown in Figure 3-3: (1) Persistent Area Assessment (PAA), (2) Mission Overwatch (MO), and (3) Situation Development (SID).

[4]Ibid.

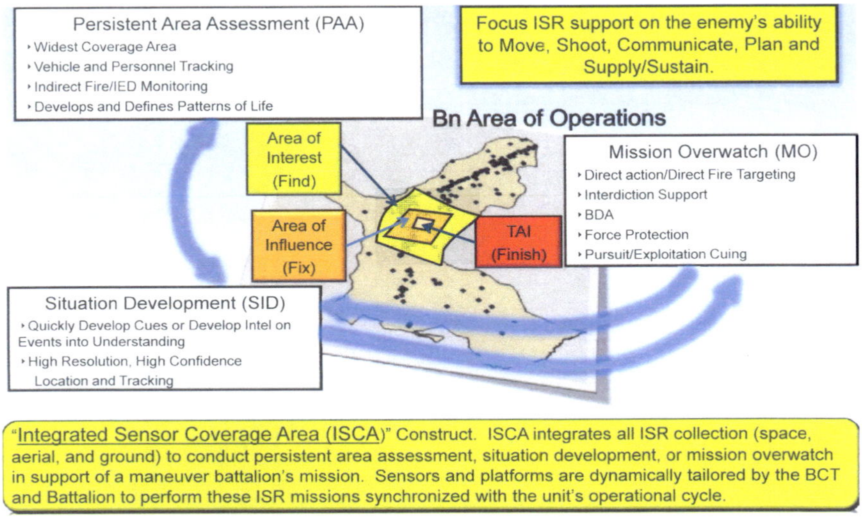

FIGURE 3-3 The Army's Integrated Sensor Coverage Area (ISCA) construct that defines three intelligence, surveillance, and reconnaissance (ISR) mission sets. SOURCE: LTG Richard Zahner, Deputy Chief of Staff, G-2, Headquarters, U.S. Army. "Military Intelligence Rebalance." Presentation to the committee, November 9, 2011.

The ISCA construct integrates all ISR collection in support of a maneuver battalion's mission, with sensors and platforms dynamically tailored by the BCT and battalion, synchronized by the unit's operational cycle. The "hourglass chart" (Figure 3-4) depicts how the ISCA construct is incorporated into the ISR capability development process, which translates the strategy into requirements and the requirements into sensors and platforms, with sensor and platform attributes that are summarized in Figure 3-5.

What is distinctive about this process is that the Army investment strategy to deliver ISR capabilities begins with a threat-and-environment-based definition of ISR requirements. These are then deconstructed into mission requirements and subsequently into three ISR mission sets, with variable sensing requirements. These sensing requirements are then translated into sensors and platforms that allow force-development options to be evaluated in a holistic, needs-based manner that is quantifiable, repeatable, transparent, and easy to explain. The selection of appropriate sensor and platform attributes provides a set of relevant and consistent criteria for defining and assessing both sensor and system performance, which, in turn, informs acquisition decisions. Characteristics applicable to airborne collection assets are summarized in Box 3-1.

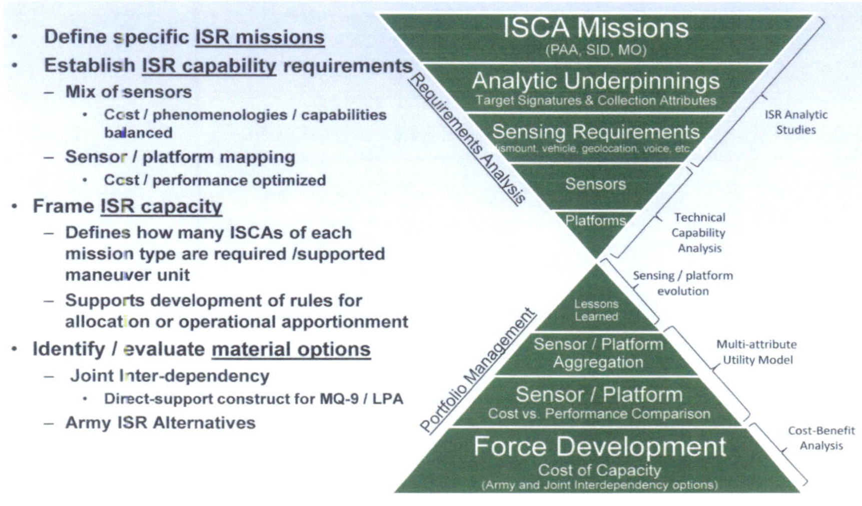

FIGURE 3-4 The Army's intelligence, surveillance, and reconnaissance (ISR) capability development process. SOURCE: LTG Richard Zahner, Deputy Chief of Staff, G-2, Headquarters, U.S. Army. "Military Intelligence Rebalance." Presentation to the committee, November 9, 2011.

Mission	Purpose	Key Tasks
Persistent Area Assessment (PAA)	Provide broad area sensing of the ISCA to develop enemy communications networks, activity, and movement	• Track vehicles • Maintain 24/7 broad area coverage over the area of operations • Support forensic and real-time exploitation • Detect personnel movement in select locations • Detect and geolocate rocket, mortar, heavy weapon explosions and flashes • Passive intercept of enemy communications
Situation Development (SID)	Receive cues from variety of sources to develop situation and target understanding	• Enable rapid change of aspect angle *or* maintenance of aspect angle to maximize coverage on specific targets and minimize track break lengths • Rapidly geolocate advanced communications devices • Support simultaneous high-resolution monitoring of several targets
Mission Overwatch (MO)	Conduct Multi-sensor ISR overwatch to current operations	• Temporal, geolocation, identity resolution • Support rapid change of aspect angles on target • High-definition/resolution monitoring • Rapidly geolocate advanced communications devices

Sensor – Platform Optimization

Threat Observables	Sensor Attributes	Platform Attributes
• Move • Shoot • Communicate • Planning • Supply/Sustainment	• Resolution • Area • Coverage • Break Length	• Altitude • Duration/Time on Station • Payload Capacity • Manned/Unmanned • Infrastructure Requirements

FIGURE 3-5 The Army's Integrated Sensor Coverage Area (ISCA) functions and intelligence, surveillance, and reconnaissance (ISR) mission requirements. SOURCE: LTG Richard Zahner, Deputy Chief of Staff, G-2, Headquarters, U.S. Army. "Military Intelligence Rebalance." Presentation to the committee, November 9, 2011.

BOX 3-1
Characteristics of the Aerial Layer Construct

1. Sensors must be optimized to support Integrated Sensor Coverage Area (ISCA)-related information collection (Intelligence, Surveillance, and Reconnaissance [ISR]) operations of Persistent Area Assessment (PAA), Situation Development (SID), and Mission Overwatch (MO).
2. The appropriate multiple-intelligence sensor array must be resident on dedicated ISR platforms to meet intelligence requirements associated with unified land operations (formerly, full-spectrum operations).
3. Intelligence sensors must be assigned (and possess the resolution requirements) to platforms that possess the endurance to support the specific requirements of the ISCA concept—PAA, SID, and MO.
4. Intelligence captured by these sensors must be accessible by forward-deployed processing, exploitation, and dissemination (PED) (in the case of Intelligence 2020 concepts, in the PED Company of the Military Intelligence [MI] Pursuit and Exploitation Battalion, the PED detachment of the MI Brigade, and the proposed PED element located in the Aerial Exploitation Battalion co-located at the Corps Headquarters). In turn these PED "platforms" are linked in the Intelligence Readiness Operations Capability (IROC) network, which will provide intelligence overwatch for deployed units as well as expand analytical and intelligence exploitation opportunities.

Finding 3-1. The U.S. Army's ISCA construct uses a process that links requirements analysis with force development and portfolio management in a way that helps synchronize planning and execution. Keys to this linkage are the ISCA analytical underpinnings and the methodology that enables sensor-platform aggregations. Additionally, the ISCA construct uses measured performance to inform acquisition decisions in a manner that lends transparency, responsiveness, and repeatability.

U.S. Navy

The overall U.S. Navy (USN) requirement-generation process is governed by Secretary of the Navy Instruction 5000.2E and defines a capabilities-based approach to developing and delivering technically sound, sustainable, and affordable military capabilities.[5] The process is implemented by means of the Naval Capabili-

[5]USN. 2011. *Department of the Navy Implementation and Operation of the Defense Acquisition System and the Joint Capabilities Integration and Development System.* September 1. Available at http://nawctsd.navair.navy.mil/Resources/Library/Acqguide/SNI5000.2E.pdf. Accessed April 13, 2012.

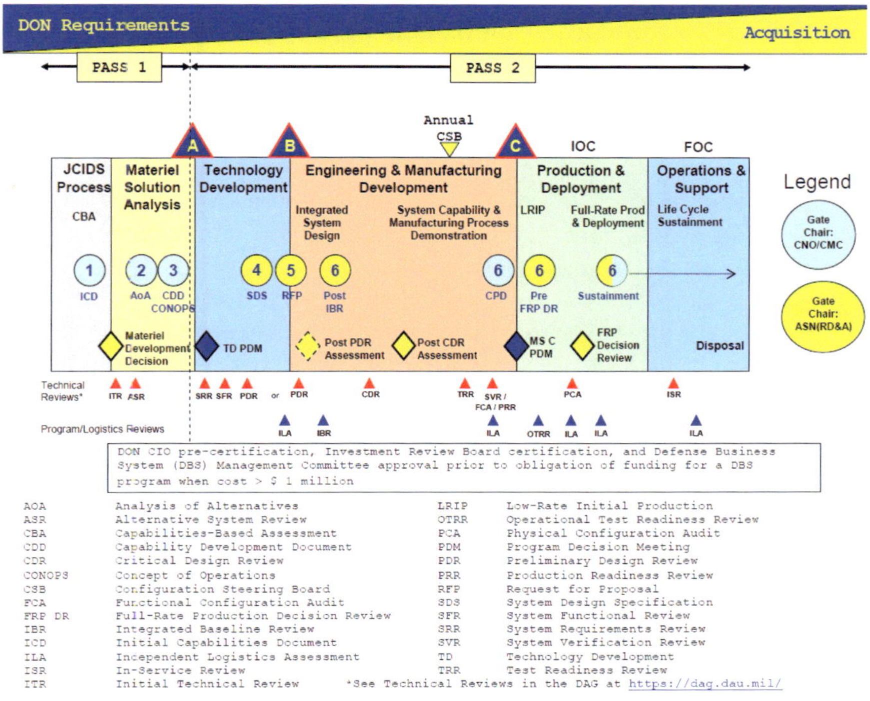

FIGURE 3-6 Department of the Navy requirements and acquisition, two-pass, six-gate process. SOURCE: Paul Siegrist, N2N6F2 ISR Capabilities Division. Personal communication to the committee, March 6, 2012.

ties Development Process (NCDP), the Expeditionary Force Development System (EFDS), and the Joint Capabilities Integration and Development System (JCIDS) to identify and prioritize capability gaps and integrated DOTMLPF solutions. The Chief of Naval Operations (CNO) is the user representative for executing actions to identify, define, validate, assess affordability determinations, and prioritize required mission capabilities through JCIDS, allocating resources to meet requirements through the Planning, Programming, Budgeting, and Execution System (PPBES). For ISR capabilities and requirements, Navy N2/6 coordinates with the Office of the CNO (N81) throughout the process.

The NCDP creates the Integrated Capabilities Plan, which translates strategic guidance and operational concepts to specific warfighting capabilities. The Navy uses two flag-level forums—the Naval Capabilities Board and the Resources and Requirements Review Board—to review and endorse all JCIDS proposals and documents. In translating requirements to operational capability, the Navy employs the two-pass, six-gate process depicted in Figure 3-6. This process ensures alignment between service-generated capability requirements and systems acquisition.

It supports Program Objective Memorandum development as well as urgent need and rapid development in streamlined, tailored implementations. A brief description of this process follows:

- *Gate 1* reviews and grants authority for the Initial Capabilities Document submission to joint review, validates the proposed analysis of alternatives (AOA) study guidance, endorses the AOA study plan, and authorizes continuation to the Material Development Decision.
- *Gate 2* reviews AOA assumptions and the total ownership cost estimate, approves the AOA preferred alternative, approves the creation of the Capabilities Development Document (CDD) and Concept of Operations (CONOPS), approves the initial Key Performance Parameters and Key System Attributes, reviews program health, and authorizes the program to proceed to Gate 3 prior to Milestone A.
- *Gate 3* approves initial CDD and CONOPS; supports the development of the service cost position; reviews technology development and system engineering plans; provides full funding certification; validates requirements traceability; considers the use of new or modified command, control, communications, computers, and intelligence (C4I) systems; reviews program health; and grants approval to continue to Milestone A.
- *Gate 4* approves a formal system development strategy and authorizes programs to proceed to Gate 5 or Milestone B.
- *Gate 5* ensures readiness for Milestone Decision Authority (MDA) approval and release of the formal Engineering and Manufacturing Development (EMD) Request for Proposal to industry, provides full funding certification, and reviews program health and risk.
- *Gate 6* follows the award of the EMD contract and satisfactory completion of the initial baseline review, assessing the overall health of the program. Reviews at Gate 6 also endorse or approve the Capabilities Production Document, assess program sufficiency and health prior to full-rate production, and evaluate sustainment throughout the program life cycle.

In summary, the application of the process employed by the Navy and applied to urgent needs involves streamlining and tailoring requirements and assessing options more rapidly than the normal process, and expediting technical, programmatic, and financial decisions as well as procurement and contracting.

Finding 3-2. The U.S. Navy's capability-based process is collaborative across the Department of the Navy and is synchronized with the PPBES and system acquisition life cycles. The process can be streamlined to address urgent needs. The process deals largely with naval requirements; utilizes existing PCPAD (planning and direction, collection, processing and exploitation, analysis and production, and dissemination)/TCPED (tasking, collecting, processing, exploitation, and dissemination) architectures; and connects with other ISR enterprise providers through the Office of the Under Secretary of Defense for Intelligence (OUSD[I]).

Office of the Under Secretary of Defense for Intelligence

The mission and vision of the OUSD(I) require an integrated approach to ISR across the Department of Defense (DoD): a global and horizontally integrated DoD intelligence capability consisting of highly qualified professionals and skilled leaders employing advanced technologies dedicated to supporting the needs of the warfighter and the Director of National Intelligence (DNI).[6] The OUSD(I) for Portfolios, Programs and Resources (PP&R) oversees the development and execution of a balanced portfolio of military and national intelligence capabilities.[7] Toward this end, the Battlespace Awareness (BA) portfolio builds the ISR investment strategy by balancing capabilities across TCPED, as shown in Figure 3-7.[8]

As shown, the OUSD(I) process leading from national-level strategy to budget decisions involves numerous organizations and staffs. Capability needs are derived from national-level defense and intelligence guidance and strategy. These needs are translated into an ISR investment strategy, with a portfolio of programs constructed and shaped to provide an optimal mix of capabilities for TCPED and analysis, given political, budgetary, and national security realities. Success depends on an understanding of top-level priorities, knowledge of ISR requirements and system capabilities, open communication (transparency), and effective collaboration among the participants.

Within the OUSD(I), the Director, Battlespace Awareness and Portfolio Assessment (BAPA) has responsibility for assessing and recommending the optimal the mix of BA capabilities to the warfighter. Figure 3-8 shows a number of key activities conducted by the BAPA staff to support portfolio development and their relationship to the PPBES process.

---

[6]DoD. 2005. "Under Secretary of Defense for Intelligence (USD(I))." Directive 5143.01. Available at http://www.fas.org/irp/DoDdir/DoD/d5143_01.pdf. Accessed February 28, 2012.

[7]DoD. 2010. "Fiscal Year 2011 Budget Estimates. Office of the Secretary of Defense (OSD)." Washington, D.C.: Department of Defense. Available at http://comptroller.defense.gov/defbudget/fy2011/budget_justification/pdfs/01_Operation_and_Maintenance/O_M_VOL_1_PARTS/OSD_FY11.pdf. Accessed February 28, 2012.

[8]Col Anthony Lombardo, Deputy Director, ISR Programs, Agency Acquisition Oversight, Office of the Under Secretary of Defense (Intelligence). "OUSD(I) Overview." Presentation to the committee, October 7, 2011.

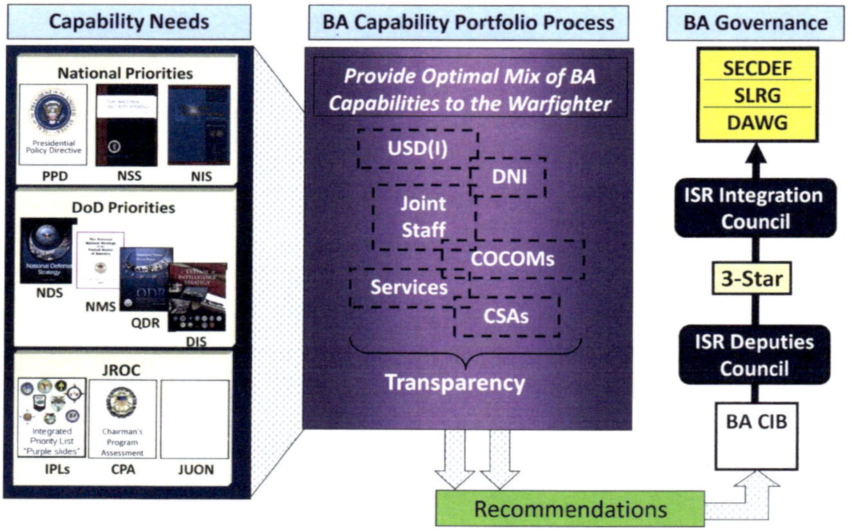

FIGURE 3-7 Office of the Under Secretary of Defense for Intelligence (OUSD[I]) strategy to budget involves numerous organizations and staffs. NOTE: Acronyms are defined in the list in the front matter. SOURCE: Col Anthony Lombardo, Deputy Director, ISR Programs, Agency Acquisition Oversight, Office of the Under Secretary of Defense (Intelligence). "OUSD(I) Overview." Presentation to the committee, October 7, 2011.

BAPA develops the ISR roadmap every 2 years, as directed by Congress. The roadmap provides information on the current ISR portfolio, its ability to meet national and defense intelligence strategies, and how the portfolio will change to remain relevant and maximize capability. Individual systems are addressed, but the overarching goal is to evaluate the portfolio and an integrated architecture. The Consolidated Intelligence Guidance gives both the DNI and OUSD(I) programmatic and budgetary guidance for programs and budgets that fall under the National Intelligence Program (NIP) and Military Intelligence Program (MIP), and it provides strategic priorities, program guidance, and areas in which to assume risk. It also directs studies when necessary to help resolve programmatic questions and uncertainties. A significant amount of analysis underpins the portfolio assessment process. The analysis comes in various forms, from major studies with cross-community participation, to Capability Area Deep Dives (CADDs), which are relatively short, intense assessments of specific issues led by the BAPA staff. Assessment efforts feed focus area teams, which are organized by domain (i.e., sea, air, and space) and help frame BA portfolio issues that need resolution.

The OUSD(I) for PP&R recognizes that current processes for prioritizing needs and analyzing risk should be improved in order to address acknowledged

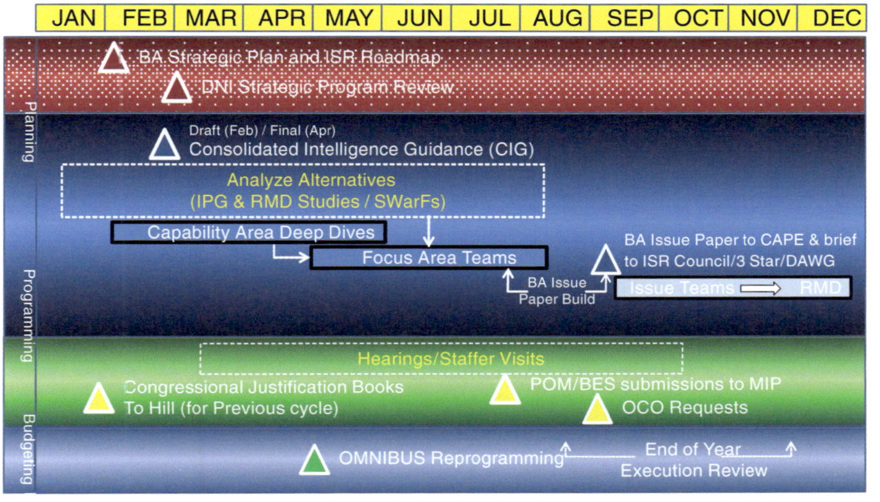

FIGURE 3-8 Office of the Under Secretary of Defense for Intelligence (OUSD[I]) and the Planning, Programming, Budgeting, and Execution System (PPBES) process. NOTE: Acronyms are defined in the list in the front matter. SOURCE: Col Anthony Lombardo, Deputy Director, ISR Programs, Agency Acquisition Oversight, Office of the Under Secretary of Defense (Intelligence). "OUSD(I) Overview." Presentation to the committee, October 7, 2011.

shortfalls, which include the following: (1) little consideration of trade-offs among cost and schedule and performance, (2) no prioritization across portfolios and little to no risk analysis, (3) the overly bureaucratic and time-consuming nature of the processes, and (4) the impact on shaping the force. The OUSD(I) for PP&R also seeks a more dynamic and iterative process throughout a program's life cycle—one that will revisit validated requirements when necessary and adjust to strategy shifts and changes in the threat, considerations that are very timely.[9] Additionally, the OUSD(I) for PP&R has recommended changes to the Joint Requirements Oversight Council (JROC) and Functional Capabilities Boards (FCBs).[10] Although OUSD(I) does not have a "standard" modeling and simulation (M&S) tool kit per se, it leverages tools and Systems Engineering and Technical Assistance (SETA) developed by Federally Funded Research and Development Centers, contractors, the services, and the Office of the Secretary of Defense (OSD) on a case-by-case basis to address specific questions (see Box 3-2). For example, the Satellite Took Kit (STK)® and the Satellite Orbit Analysis Program (SOAP) have been used primarily to help leadership and decision makers visualize overhead ISR systems and evaluate

[9]Ibid.

[10]Ibid.

BOX 3-2
Intelligence, Surveillance, and Reconnaissance (ISR) Tools Used by the Office of the Under Secretary of Defense (Intelligence) (OUSD[I])

Satellite Tool Kit (STK)®. A three-dimensional-visualization tool that can display orbit geometries of space systems and realistic views of airborne and terrestrial assets as well. Used primarily to facilitate the understanding of ISR satellite capabilities and limitations, notably persistence and area coverage.

Satellite Orbit Analysis Program (SOAP). An interactive, three-dimensional orbit visualization and analysis program that can generate an unlimited number of world, XY plot, and textual views. Used primarily to show persistence and coverage of overhead systems and to assist with specific engineering assessments of overhead systems. (Note: Used by the National Reconnaissance Office [NRO] on its NRO Management Information System [NMIS] terminals).

DyCAST (Aerospace). A relay satellite communications scheduling and analysis tool that helps resolve contention and perform optimal communications resource allocation. It was used to support the Airborne Intelligence Surveillance and Reconnaissance (AISR) Analysis of Alternatives and several communications studies sponsored by the Department of Defense and the intelligence community.

Communications Architecture Systems Assessor (CASA). Developed by Aerospace Corporation to compare the performance of alternative communications architectures under different operational (dynamic) scenarios. Has become the "tool of choice" of OUSD(I) for evaluating communications sufficiency and has also been used by NASA and other intelligence community entities to investigate communications issues.

Joint Force Operational Readiness Combat Effectiveness Simulator (JFORCES). A government-owned simulation tool capable of producing operationally credible data on the interactive behavior of sensors, command and control, weapons and communications systems for both friendly and opposing forces. JFORCES has supported numerous engineering studies as well as command, control, communications, computers, intelligence, surveillance and reconnaissance (C4ISR) architecture assessments.

Boeing Engineering Analysis Support Tool (BEAST). A high-fidelity aerospace system simulator for subsystem, system, and system-of-systems analysis.

Architecture Evaluation Tool (AET). Developed by National Security Space Office (NSSO) (now Executive Agent for Space [EA4S] staff); displays aggregated capabilities and metrics, such as measures of effectiveness, for various architectures as a function of cost. It allows the decision maker to interactively adjust the weighting of various metrics and gauge the sensitivity of selected architecture capabilities to cost.

persistence and coverage issues. The remaining tools have been used most often to assess communications sufficiency under different scenarios and in the face of expanding collection platforms, especially airborne ISR.

In many respects, the most common analysis approach used by OUSD(I) and the one most commonly observed in CADDs is what can be described as a straightforward empirical analysis approach based on operational data. This approach has dominated in recent years because most of the leadership focus has been on ISR performance in Iraq and Afghanistan, for which empirical data can be obtained. As the fight draws down and the focus shifts to longer-range architectural issues, more M&S tools will likely be used.

> **Finding 3-3.** The CP&A-like process employed by OUSD(I) addresses ISR enterprise concerns across the DoD and the IC and includes consideration of the capabilities of enterprise networks and PCPAD and TCPED. The OUSD(I) recognizes the need to improve the capability development process in the following ways: (1) by attaining better up-front fidelity on trade-offs involving cost and schedule and performance, (2) by providing more analytic rigor and risk/portfolio analysis, (3) by placing stronger emphasis on prioritizing requirements and capabilities, and (4) by strengthening alignment of the acquisition process.

EXAMPLES OF INDUSTRY PROCESSES FOR PROVIDING CAPABILITY PLANNING AND ANALYSIS

In addition to pertinent government approaches to CP&A, select industry CP&A-like processes and their associated tools were reviewed for their potential applicability to the Air Force CP&A process. Although many of the processes presented use a similar high-level approach that involves a requirements and needs analysis, a capabilities gap analysis, and a solutions analysis, the levels of detail, complexity, and development and employment of tools varies considerably among industry processes. Ultimately, the output of the efforts is generally a report and/or a brief that presents alternatives in terms of priorities, cost, mission utility, and risk.

Booz Allen Hamilton

Booz Allen Hamilton (BAH) provides SETA services to the U.S. government, as well as consulting services to the ISR community.[11] BAH advocates Capabilities-

[11]Information on BAH's systems engineering and integration efforts is available at http://www.boozallen.com/consulting/engineer-operations/systems-engineering-integration. Accessed February 28, 2012.

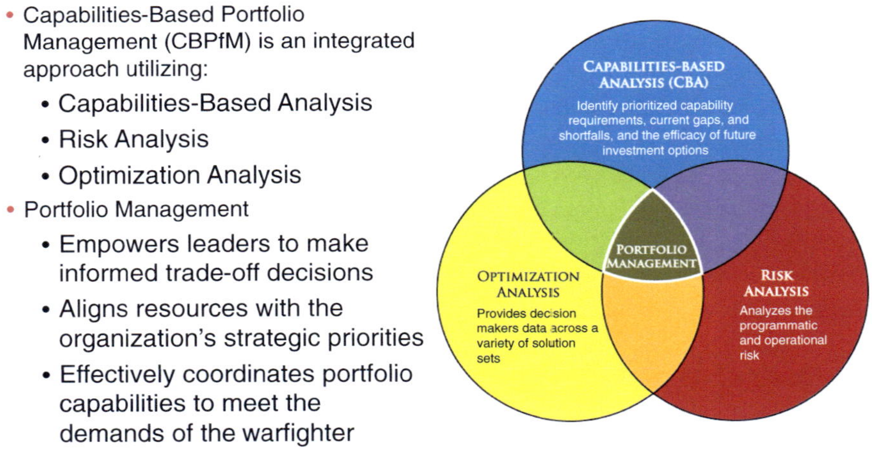

FIGURE 3-9 Booz Allen Hamilton's Capabilities-Based Portfolio Management (CBPfM) process. SOURCE: Scott Gooch, Principal, and Christopher Anderson, Lead Associate, Booz Allen Hamilton. "Capabilities-Based Portfolio Management: Methods, Processes, and Tools." Presentation to the committee, January 5, 2012.

Based Portfolio Management (CBPfM) that combines capabilities-based analysis, risk analysis, and optimization analysis to inform decision makers conducting portfolio management.[12] BAH capabilities-based analysis tools can be applied across different parts of the trade-off study process, depending on how both the process and the tool are customized for a particular portfolio or government customer. The benefits ascribed to this approach, as shown in Figure 3-9, are as follows: CBPfM "empowers leaders to make informed trade-off decisions, aligns resources with the organization's strategic priorities, and effectively coordinates portfolio capabilities to meet the demands of the warfighter."[13] The CBPfM process, presented in Figure 3-10, follows the basic industry flow of needs analysis, gap analysis, and solution analysis, and includes cost, schedule, performance, and risk analyses.

Within the CBPfM process, BAH utilizes a number of commercial off-the-shelf (COTS) software systems that ingest different types of data (budget, capability, risk) and optimize and prioritize alternatives across user-defined objectives and constraints. In addition, a number of the tools employed by BAH facilitate the

[12]Scott Gooch, Principal, and Christopher Anderson, Lead Associate, Booz Allen Hamilton. "Capabilities-Based Portfolio Management: Methods, Processes, and Tools." Presentation to the committee, January 5, 2012.

[13]Ibid.

Step	Key Questions	Potential Issues	Lessons Learned
1 Define and Scope the Portfolio	▯ What capabilities / programs are part of the ISR portfolio? ▯ What is the relationship to others (e.g., National Intelligence Program)?	▯ Must balance the broad and complex nature of enterprise-level portfolio with the practical aspects of exactly which data are helpful	▯ Leadership needs to be a champion ▯ Define scope from the inside-out of the organization
2 Establish and Measure Portfolio Objectives	▯ How do current and planned investments align with Warfighter mission needs?	▯ Can be difficult to quantify contributions across domains and degrees of alignment	▯ Link the portfolio framework to an authoritative framework (i.e., JCAs) for traceability and defensibility
3 Conduct Portfolio Risk Analysis	▯ What are the risk levels, sources of risk, and types of risk within each of the programs and to the portfolio? ▯ How do portfolio changes affect operational risk?	▯ Need to standardize risk measurement across portfolio ▯ Risk impacts for enterprise portfolios can be difficult to quantify	▯ Relate risk to cost, schedule, and performance impacts ▯ Make risk reporting more objective and standardized wherever helpful
4 Measure Response to Funding Changes	▯ What are the implications for funding changes to cost, schedule, performance, and risk?	▯ Data can be very sensitive ▯ Difficult to predict how various portfolio elements would react to funding changes	▯ Avoid schedule-slip only mentality ▯ Present this as an opportunity to make defensible cases about the impact of funding changes
5 Determine Portfolio Constraints	▯ What constraints are necessary to reflect portfolio realities / preferences? ▯ What are the interdependencies across investments?	▯ Increasingly complex portfolios and their interdependencies may require detailed constraints	▯ Review assumptions at each step of portfolio analysis ▯ Ensure external influences and funding constraints are captured
6 Analyze Portfolio	▯ What capability can be achieved given certain funding? ▯ If forced to make cuts, in which areas can risk be accepted?	▯ Portfolio analysis must be traceable and repeatable to quickly respond to a variety of whatif drills	▯ Account for both proficiency and sufficiency in analysis ▯ Populate data with distributions instead of "best-guess" points
7 Conduct Sensitivity Analysis	▯ Is the optimized solution feasible and sensitive to any inputs?	▯ Enterprise analysis may introduce many variables and assumptions	▯ Use sensitivity analysis to identify areas for further review not automatic course correction

*Steps are shown as strictly linear for display purposes only

FIGURE 3-10 The Capabilities-Based Portfolio Management (CBPfM) process: steps and intial discussion points. SOURCE: Scott Gooch, Principal, and Christopher Anderson, Lead Associate, Booz Allen Hamilton. "Capabilities-Based Portfolio Management: Methods, Processes, and Tools." Presentation to the committee, January 5, 2012.

real-time visualization of portfolio performance over time and support "what-if" drills with schedule slips (see Table 3-1).[14]

Finding 3-4. Booz Allen Hamilton's Capabilities-Based Portfolio Management process requires leadership engagement, diverse skill sets to analyze a portfolio, and stakeholder participation and transparency. The resultant assessments are repeatable and rigorous enough to enable long-term planning, yet agile enough to incorporate new scenarios, priorities, and missions. The process includes modeling of extant TCPED and communications architectures, which yields more realistic estimates of cost and performance and risk. Although many results are scalable, any consideration of broader, more complex enterprises requires good analytical judgment for the development of the right approach.

TASC

TASC provides systems engineering and integration solutions to the DoD, the intelligence community, and civil agencies. TASC illustrated its Capability-Based Assessment (CBA) process through Multi-Resolution Analysis (MRA) applying

[14] Ibid.

TABLE 3-1 Sample of Tools Employed by Booz Allen Hamilton for Facilitating Real-Time Visualization of Portfolio Performance Over Time and Supporting "What-If" Drills with Schedule Slips

Tool Name	Category	Tool Product	Benefits
Simio® Simio Discrete-Event Simulation Software *(Commercial off-the-shelf)*	Mission Utility	Simulation	Higher-fidelity performance analysis.
Satellite Tool Kit (STK)® *(Commercial off-the-shelf)*	Physics-Based Capability	Modeling and Analysis Software	Higher-fidelity performance analysis.
EADSIM Extended Air Defense Simulation *(Government owned)*		Simulation	Higher-fidelity performance analysis.
ISR FOCUS *(Booz Allen Hamilton)*	Integrated Decision Aides	Simulation: Tool product uses commercially available simulation such as Satellite Tool Kit® and Simio®. The product itself is currently an Excel-based dashboard built without Visual Basic for Applications (VBA). It can be deployed rapidly to any network.	The repeatable FOCUS assessment process enables managers and leaders to "see."
Advanced Interactive Multidimensional Modeling Simulation (AIMMS)® *(Commercial off-the-shelf)*		Optimization	Repeatable, traceable optimization and intuitive dashboards.
Expert Choice® Decision Support Software *(Commercial off-the-shelf)*		Decision Aides	Clearly solicits priorities.
Resource Allocation Model (RAM) *(Booz Allen Hamilton)*		Portfolio Management	Aligns programs and capabilities to strategy and guidance. Establishes standardized metrics to compare programs. Streamlines inputs, processes, and outputs. Tailorable to one's organization.
Dynamic Capability Assessment Model (DCAM) *(Booz Allen Hamilton)*		Portfolio Management	Provides a unique dashboard visualization of cost, schedule, and performance in an interactive and dynamic environment.
Decision Lens® Analysis *(Commercial off-the-shelf)*	Financial and Business Analysis	Financial and Business Analysis	Embedded prioritization and solver tools.

NOTE: For a more complete list of tools used by both government and industry, see Appendix C in this report.
SOURCE: Booz Allen Hamilton. 2012. Written communication. Response to inquiry from the committee.

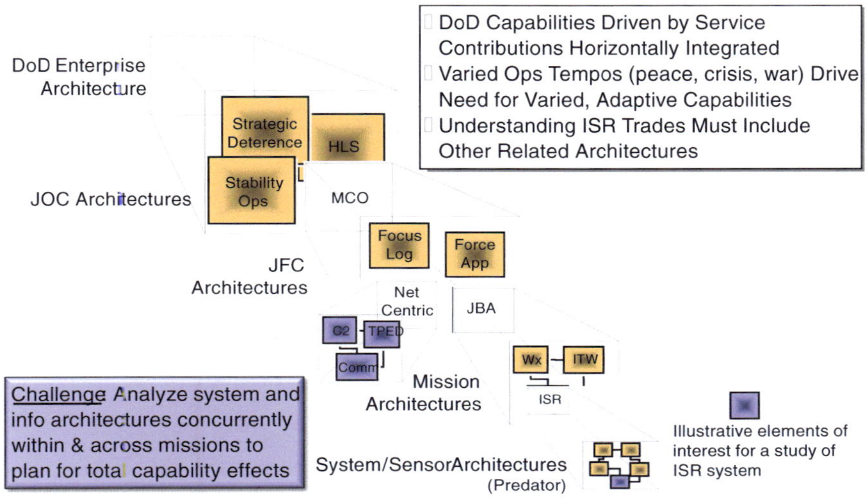

FIGURE 3-11 TASC's solution to analyzing complex domains begins with a layered, iterative approach, segregating and describing architectures within and across missions. NOTE: Acronyms are defined in the list in the front matter. SOURCE: Doug Owens, Manager, Enterprise Analysis, Defense Business Unit, TASC. "An Enterprise Approach to Capability-Based Analysis: Best Practices, Tools, and Results." Presentation to the committee, January 5, 2012.

various COTS/government off-the-shelf (GOTS) and TASC tools to the analysis of large-scale, complex domains.[15] The TASC solution to analyzing complex domains begins with a layered, iterative approach, segregating and describing architectures within and across missions, as shown in Figure 3-11. In this way, TASC described a quantifiable analysis across complex domains, informed by affordability, and with traceability from requirements to decision outcomes.

The basic premise of the TASC approach is that complex domains of capability can be analyzed from different perspectives with tailored models and tools appropriate for each perspective, and the various segments of the analysis are integrated to provide traceability of cause and effect for the combined total impact, shown in Figure 3-12.[16] For the ISR mission area, those perspectives include the following: sensor and collection platform performance; the network topology connectivity that enables the overall ISR mission; the command and control of the various assets; the communications capabilities and allocations; the vulnerabilities of the information architecture for the command, control, communications, and computers (C4) capabilities that enable ISR; the processes for TCPED information in

[15] Doug Owens, Manager, Enterprise Analysis, Defense Business Unit, TASC. "An Enterprise Approach to Capability-Based Analysis: Best Practices, Tools, and Results." Presentation to the committee, January 5, 2012.

[16] Ibid.

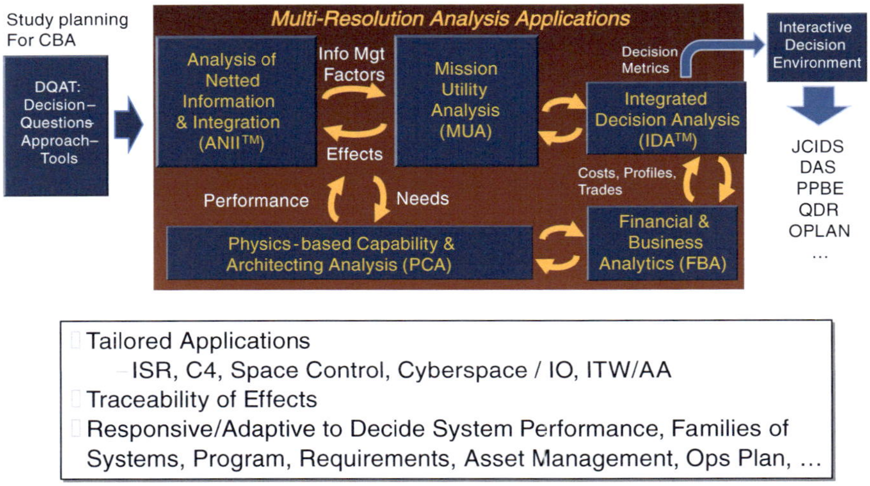

FIGURE 3-12 TASC's approach to Multi-Resolution Analysis (MRA) integrates various perspectives. SOURCE: Doug Owens, Manager, Enterprise Analysis, Defense Business Unit, TASC. "An Enterprise Approach to Capability-Based Analysis: Best Practices, Tools, and Results." Presentation to the committee, January 5, 2012.

support of operations; and the manpower and infrastructure through which the ISR missions are accomplished.[17] MRA enables the examination of each of these elements within an integrated capability context using an interactive, iterative flow through the analysis. Multi-criteria methods are then correlated to cost estimating and program risk analysis, cost profiling, organization assessments, and six-sigma process improvement. The use of full-spectrum analytics within an integrated, interactive process combines the science of systems engineering and systems integrations with decision making.

TASC executes full-spectrum, cross-domain, multi-resolution analyses using a variety of GOTS, COTS, and custom tools to address the command, control, communications, computers, intelligence, surveillance, and reconnaissance (C4ISR) enterprise. For example, GOTS/COTS tools like ADIT (Advanced Data Integration Toolkit) for multi-intelligence data fusion analysis, GeoViz for geolocation performance analysis, STK for tracks and orbital coverage analysis, SEAS (System Effectiveness Analysis Simulation) for ISR mission effects and CONOPS development, JIMM (Joint Integrated Mission Model) for integrated operations analysis and detailed constructive analysis, and other tools provide the physics-based analysis for the quantification of capability impacts.

A sample of TASC custom tools and processes for ISR capability analysis is shown in Table 3-2. Among them are ANII™ and a-MIND™, which allow the

[17]Ibid.

TABLE 3-2 Sample of Tools, with Products and Benefits of Each, Employed by TASC

Tool Name	Category	Tool Product	Benefits
JFORCES: Joint Force Operational Readiness Combat Effectiveness Simulator *(TASC)*	Mission Utility	Stochastic and deterministic simulation	Provides robust analysis of total capability. Archives every element of a simulation to allow analysts to create new metrics to explore issues without re-executing simulation.
a-MIND™ with ANII™ Process: automated Mission Impact of Network Design *(TASC)*	Processing, Exploitation, Dissemination (PED): Analysis Communications Data Integration Decryption Language Translation Data Reduction	Statistical relational models; mission impact of network design, cyber impacts on mission effectiveness; cyber mitigation options	Reduces analysis cycle time by means of rapid diagnostic evaluation of network or architecture for mission impacts to identify alternative structures from potentially millions of options. Quantifies correlation of networks to missions.
TPAT: TCPED Process Assessment Tool (built in ExtendSIM) *(TASC)*		General simulation	Visualization of process provides quick reference point to identify bottlenecks and inefficiencies. Enables rapid exploration of process options.
CACI: Collection Architecture Capability Influence *(TASC)*		Inference modeling (Infer potential changes in outcome or effects from possible variations in metric results)	Reduces architecture costs by identifying key elements and indifferent elements, allowing capability development and selection to focus on critical pieces.
Tasking to Value (T2V): Geospatial Modeling Environment *(TASC)*		Simulation	Increased operational reality and ability to assess ISR operations effectiveness.

continued

TABLE 3-2 Continued

Tool Name	Category	Tool Product	Benefits
SMART: Strategic Multi-Attribute Resource Tool *(TASC)*	Integrated Decision Aides	Multi-attribute Utility Analysis	Reduces decision complexities and provides real-time decision-maker interaction with metric data to explore trade space of options.
QATO: Quick Automated Tool for Optimization *(TASC)*		Excel-based statistical model	Rapid correlation of metrics and cost profiles, reducing programming trades and impact analysis of program decisions.
MIATI: Multi-theater Integrated Allocation Tool for ISR *(TASC)*		Asset allocation	Reduces time and cost of analysis by quickly narrowing the trade space of asset management.
H-BEAM with MESA Process Horse Blanket Enterprise Architecture Methodology (TASC)	Architecture Analysis	Technical performance metrics of system and family of systems effectiveness. Assessment of system impacts to mission effects as scoping analysis for subsequent detailed Mission Utility Analysis (MUA).	Consolidated display of architecture effects and contributing elements.
CERA: Cost Estimating and Risk Analysis Process *(TASC)*	Financial and Business Analytics	System and family of systems cost estimates and profiles; correlation of costs to metric performance from PCA, ANII™, or MUA	Develop credible estimates and profiles for systems and families of capability. Assess risks of programmatic changes on capability effects.

NOTE: For a more complete list of tools used by both government and industry, see Appendix C in this report.
SOURCE: TASC. 2012. Written communication. Industry and Government ISR Tools and Processes. Response to inquiry from the committee.

analysis of interconnected capabilities and associated cyberspace vulnerabilities in a C4ISR information architecture; Strategic Multi-Attribute Research Tool (SMART), which supports metric-driven decision analysis, including uncertainty and cost-benefit trade-offs; Collection Architecture Capability Influence (CACI), which supports risk analysis in collection architectures; and the Mission Engineering and Systems Analysis (MESA) process paired with the Horse Blanket Enterprise Architecture Methodology and visualization tool (H-BEAM), which graphically traces capability across an enterprise architecture, from strategic guidance and requirements, to systems, to architecture options, to capability impacts.

In summary, TASC described to the committee the inherent challenge in trying to analyze system and information architectures concurrently within and across missions to plan for total capability effects. Specifically, the networked architectures are extremely complex, and the TASC solution is a layered analytic discipline to provide quantifiable analysis informed by affordability. TASC maintains that MRA manages this complexity while maintaining traceability of effects through engineering analysis, family of systems and architecture trade-offs, networked information and integrated C4 for ISR, mission utility effects, and decision and costing analysis. Further, MRA provides multiple views for decisions on system technical performance parameters, network connectivity and information vulnerabilities, family of capabilities, concepts of operations, policy, total capability versus cost trade-offs, operations planning, and asset allocation.[18]

Finding 3-5. TASC's capability-based assessment process employs MRA, which in turn allows the complexity of ISR to be handled in a straightforward, transparent, tailorable, scalable, repeatable manner, incorporating a suite of tools that are optimized for a specific purpose. Such an approach can support a wide range of decisions and decision time lines.

RadiantBlue, Inc.

RadiantBlue, Inc., is a specialized provider of information technology development, consulting, and program support services for the DoD and the intelligence community.[19] As with TASC, RadiantBlue implements the mission utility analysis and physics-based capability and architecture assessment phases of an MRA process using its "Blue Sim" Tool, an ISR high-fidelity simulator with agile software that easily accommodates new assets, payloads, and requirements scenarios. Figure 3-13 illustrates a typical BlueSim model with various payload types and relevant vehicle subsystems for the ISR trade space. RadiantBlue has used BlueSim to performed detailed analysis, including analysis in the following areas: space and air, sensor performance, flight profiling, attitude and orbitology, communications, TPED (tasking, processing, exploitation, and dissemination), collection satisfaction, force sizing, architecture, and visualization. The BlueSim simulator allows the integrated analysis of space, air, and ground systems—across integrated IMINT and SIGINT payloads, with cyberspace effects, against classic portfolios of ISR targets, or target decks, and vetted DoD scenarios.

[18]Ibid.

[19]More information on RadiantBlue's mission is available at http://www.radiantblue.com/about/. Accessed February 28, 2012.

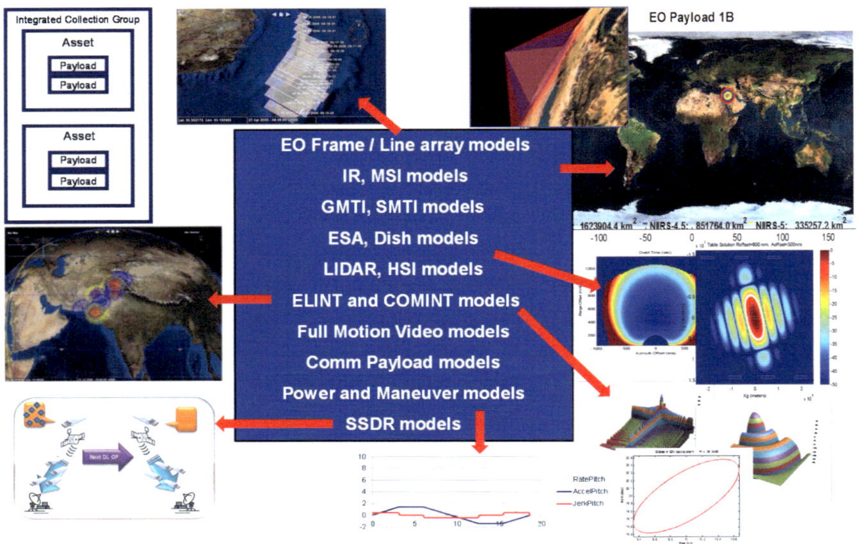

FIGURE 3-13 RadiantBlue's BlueSim accurately models the full depth of key subsystem and payload aspects of intelligence, surveillance, and reconnaissance (ISR) collection. NOTE: Acronyms are defined in the list in the front matter. SOURCE: Larry Shand, President, RadiantBlue, Inc. "RadiantBlue Modeling and Simulation Capabilities." Presentation to the committee, January 5, 2012.

Government users have employed BlueSim across the full range of analysis, from large architecture studies to detailed collection planning studies. Likewise, industry users have employed BlueSim for diverse applications, including ISR architectures, system performance, predictive simulations, and detailed system and payload studies. Table 3-3 describes the tools employed by RadiantBlue Tool Set as well as its products and benefit.

The following sections describe RadiantBlue's study process (shown in Figure 3-14).[20]

Essential Study Documents in Place, Study Kickoff, Study Trade-Space Definition

The first three steps illustrated in the RadiantBlue process description define the details of the requested study, including study tasks, the study trade space, and desired outcomes, as well as a set of study messages and themes to guide the development of study output products. The trade space details the systems, payload variations, architectures, CONOPS, target decks, and analysis vignettes simulated

[20]RadiantBlue. February 7, 2012. Written communication to the committee.

TABLE 3-3 Description of Tools Employed by RadiantBlue Tool Set, with Products and Benefits

Tool Name	Category	Tool Product	Benefits
BlueSim	Mission Utility	Simulation	Complete ISR mission effectiveness and utility. Measured in terms of classic ISR mission utility parameters (points and areas per day) or lower-level military/tactical utility parameters such as number or percentage of key enemy behavior detected.
BlueSim and supporting payload phenomenology tools	Physics-Based Capability	Simulation	Highly detailed physics-based model that simulates key ISR system and architecture features. These supporting models provide detailed inputs to BlueSim to enable mission- and architecture-level analysis with rigorous physics-based underpinnings.
BlueSim	Architecture Analysis	Simulation	Combines large architecture analysis and accurate physics-based modeling. Unique features: integrated cross-system tipping and cueing, the ability to model graduations between unified and stovepiped tasking systems, integrated dynamic tasking, and the ability to model intelligence gained through collection (information model).
BlueSim Ground Model	Processing, Exploitation, Dissemination (PED): Analysis Communications Data Integration Decryption Language Translation Data Reduction	Simulation	Enables PED analysis either combined with or separate from the ISR collection modeling, includes features to model cyber and input/output (IO) effects such as intermittent outages, random packet loss, link degradation, node failure, and automatic communications. Includes all of the key ISR wideband network node types to include routing, processing, storage area network (SAN) storage, exploitation, etc.

NOTE: For a more complete list of tools used by both government and industry, see Appendix C in this report.
SOURCE: RadiantBlue. 2012. Written communication. Response to inquiry from the committee.

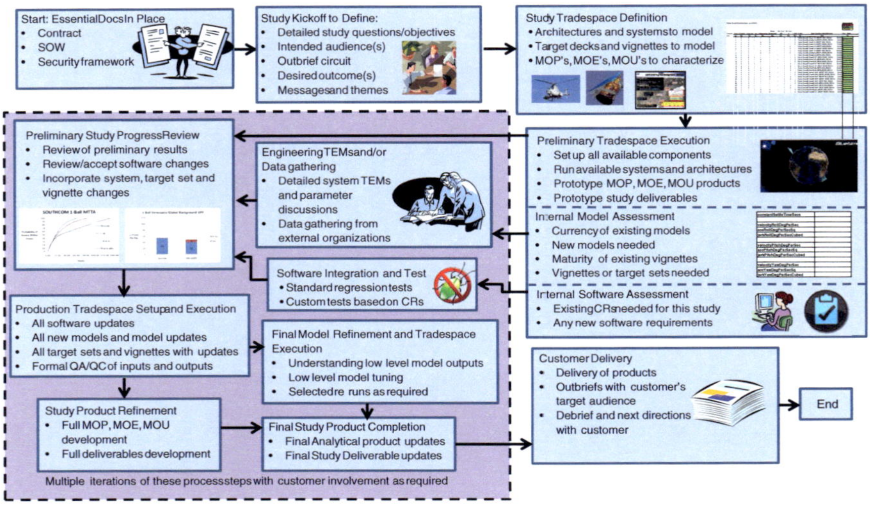

FIGURE 3-14 RadiantBlue process description. SOURCE: RadiantBlue. 2012. Written communication to the committee.

in various combinations to create the raw quantitative data that will serve as the basis of the technical analysis to be conducted in the study. Also part of the detailed study definition are the measures of performance (MOPs), measures of effectiveness (MOEs), and measures of utility (MOUs) that are used to quantify trade-space performance and allow analysis thread comparisons as the study matures.

Preliminary Trade-Space Execution, Internal Model Assessment, Internal Software Assessment

With the study trade space fully defined, RadiantBlue then uses an extensive pre-existing library of ISR system models, target decks, and vignettes to allow a study team to take existing, simulator-ready data and build significant portions of a trade space to begin runs immediately. These three parallel activities (trade space, model, and software assessments) provide analytical products that drive the next several process steps in an iterative and collaborative manner with the larger study team.

Engineering Technical Exchange Meetings, Software Integration and Test, and Progress Review

Each of the analytical products is subjected to internal model assessments and internal software reviews in which the preliminary trade-space run data results are reviewed, and a new version of the simulator is provided. Then a progress review is conducted with the prime contractor or government point of contact (POC) during which the preliminary trade-space results are discussed, sample products are reviewed, and the status with respect to model additions and changes is reviewed. This is also an opportunity for the customer POC to refine or alter the direction of the study on the basis of these preliminary results and/or programmatic, financial, or political developments outside the study team. At the end of this process step, the study team is put on a refined vector for where to take the study in terms of priorities and trade-space definition.

Production Trade Space, Study Products, Final Study Products

With the refined study direction vector, updated executable(s), and updated models, the study team then sets up the full production trade space that includes all of the key models, target decks, vignettes, CONOPS, and software features. As the simulation data emerge from the trade-space runs, the MOP, MOE, and MOU products can be developed to address the messages and themes that were defined in the study kickoff phase. Technical measures can be refined or replaced as needed, and these then feed modifications to the messages and themes as required. In conjunction with refining technical measures, detailed low-level analysis of the simulator output data is conducted to make sure that the macro-level trends that are emerging are supported by coherent physics and technology-based micro-level system behaviors. The final study products are then formulated to meet the desired study outbriefing plan and the internal needs of the study team. These final study products are typically developed collaboratively and iteratively with the primary prime contractor or government POC and other key members of the larger study team as required.

Summary

RadiantBlue works collaboratively and iteratively with the customer to refine the details of the desired study.[21] Continued collaboration with customer subject-matter experts serves to detail and enhance the understanding of the trade space, including systems, payload variations, architectures, CONOPS, target decks, analysis vignettes, MOPs, MOEs, and MOUs. With the trade space clearly defined, RadiantBlue conducts analysis of the trade space through multiple simulation runs of scenarios and vignettes. Iterative sessions between the customer and RadiantBlue serve to refine tools and scenarios, ultimately leading to study results and

[21] RadiantBlue. February 7, 2012. Written communication to the committee.

completion.[22] Finally, RadiantBlue's process (using the BlueSim simulator) requires iterative customer engagement and collaboration between operators and analysts and is supported by a large, pre-existing, model library of air and space systems.

RadiantBlue provided for the commiteee a second industry example of how a very complex set of assets and vignettes can be evaluated iteratively through an MRA process that is thoroughly documented for transparency, accuracy, and repeatability and can be tailored and scaled to customer desires.[23]

Both TASC and RadiantBlue identified analysis approaches that are responsive to their customers' needs by taking full consideration of ISR assets and trade-offs across the enterprise, spanning air, space, and, to a lesser extent, cyber effects. What is most helpful is the approach of pairing physics-based, layered analysis tools, cost-estimating, risk analysis trade-offs, along with the cost projections over various planning horizons (e.g., Analysis of Alternatives and Program Objective Memorandums) when implementing full-spectrum MRA.

> **Finding 3-6.** RadiantBlue's modeling, simulation, and analysis capability focuses on the physics-based capability and architecture analysis and mission utility analysis found in MRA. The BlueSim tool, combined with RadiantBlue's methodology, has been used to successfully support trade-space studies of various ISR and processing, exploitation, and dissemination (PED) architectures.

CONCLUDING THOUGHTS

In the committee's reviewing of the government and industry CP&A-like processes described in this chapter, it became apparent that multiple tools, including both commercial-off-the-shelf and proprietary tools, are utilized effectively across government and industry for modeling, simulation, and analysis, and that "one size does not fit all." Second, none of the non-Air Force CP&A-like processes reviewed adequately addresses the emergent challenges posed by the cyberspace domain. Third, most of the non-Air Force CP&A-like processes reviewed do not adequately deal with the complexity of PCPAD, which, in turn, can affect cost, performance, and schedule. This latter issue can also result in capabilities that are not end to end and contributes to information and data that cannot be shared, correlated, or fused by users or customers. Finally, the objective of considering a wide range of government and industry CP&A-like processes was to gain insight into potential best practices to incorporate into this study's overall recommendations. Table 3-4 maps findings to these best practices.

[22]More information on RadiantBlue's methodology is available at http://www.radiantblue.com/solutions/software-development/. Accessed February 28, 2012.

[23]Congressional professional staff members who spoke with the committee identified RadiantBlue as the best modeling organization at the architecture level.

TABLE 3-4 Best Practices and Corresponding Findings

Best Practice	Finding
Process includes consideration of "enterprise" ISR systems and/or capability.	Findings 3-3, 3-4, 3-5, and 3-6
Process is transparent, responsive, scalable, and repeatable.	Findings 3-1, 3-4, 3-5, and 3-6
Process is underpinned by multi-resolution-like analysis, modeling and simulation.	Findings 3-4, 3-5, and 3-6
Process is collaborative and links planning, acquisition, and operations.	Findings 3-1 and 3-2
Process is informed by operational metrics.	Finding 3-1
Process incorporates network/PCPAD/TCPED architectures and cyberspace considerations.	Findings 3-2, 3-3, 3-4, 3-5, and 3-6

NOTE: Acronyms are defined in the list in the front matter.

4

Toward an Enhanced Air Force Intelligence, Surveillance, and Reconnaissance Capability Planning and Analysis Process

INTRODUCTION

The objective of Chapter 4 is to propose recommendations designed to guide the Air Force toward a new intelligence, surveillance, and reconnaissance (ISR) Capability Planning and Analysis (CP&A) process that enhances rather than replaces the current process. Chapter 4 begins by presenting three overall recommendations associated with the proposed process. The chapter then presents a set of desired attributes for this process that are based on the strengths and shortcomings of the current process, as well as on best practices provided by government and industry. Lastly, the chapter describes in detail a proposed ISR CP&A process that employs the best practices identified in Chapter 3 for overcoming the shortfalls with the current process as described in Chapter 2.

RECOMMENDATIONS

Chapter 2 describes the current process used by the Deputy Chief of Staff of the Air Force for ISR (AF/A2) to plan for and assess ISR capabilities. It concludes by summarizing strengths and shortfalls of the current process garnered from interactions with various ISR and best-practice stakeholders. Chapter 3 identifies several best practices associated with government and industry CP&A processes that suggest solutions for the shortfalls identified in Chapter 2. After considering best practices in the context of strengths and shortfalls, the committee arrived at three major recommendations designed to guide the Air Force toward a new and more comprehensive ISR CP&A process. This section presents and provides a rationale for each recommendation.

Recommendation 4-1. The Air Force should adopt an ISR CP&A process that incorporates the following attributes:

- Encompasses all ISR missions;
- Addresses all ISR domains and sources, including non-traditional ISR;
- Includes all ISR assets in a sensor-to-user chain (e.g., PCPAD and communications);
- Collaborates with ISR-related entities;
- Provides traceability from process inputs to outputs;
- Is mission/scenario-based;
- Is repeatable and enduring;
- Supports trade-off analyses;
- Is scalable in size, time, and resolution; and
- Reduces labor and cost over time.[1]

Rationale: The Air Force currently has a reasonable ISR CP&A process but has indicated that this process requires improvements. The committee identified gaps in capability in the existing Air Force process, explored best practices for CP&A in both government and some industry organizations, and developed a set of desired attributes from its analysis of gaps and best practices that represent a robust ISR CP&A process. With the addition of three capabilities, the Air Force can attain the desired attributes by enhancing rather than replacing the current process. The three capabilities are as follows: (1) a front-end Problem Definition and Approach (PDA) capability, (2) a robust Multi-resolution Gap Analysis (MGA) capability, and (3) a suite of automated tools that underpin that analysis of the cost, risk, and utility associated with investment alternatives. The full process and an explanation of how it satisfies the desired attributes are presented later in this chapter. The committee acknowledges that the proposed process should accommodate the use of all levels of classified material in the analysis. Although industry presentations included examples of tools being used to process classified information, both security and

[1]The committee acknowledges that any process needs to accommodate the use of all levels of classified material in the analysis. However, security and time constraints precluded the committee from making recommendations for multi-level security analysis. Chapters 2 and 4 provide supporting discussions.

time constraints precluded the committee from making detailed recommendations regarding analyses involving multi-level security.

Recommendation 4-2. The Air Force should evolve its ISR CP&A process to an integrated, overarching ISR investment process with clear organizational responsibility identified for each subprocess.

Rationale: One of the most important actions that the Air Force can take is to implement an integrated ISR CP&A process. As described in Chapter 2, the Air Force has evolved into a current situation that has multiple, overlapping investment processes that appear to duplicate effort. The integrated process should have clearly defined roles and responsibilities for participants and clear identification of the lead for each portion of the process. Although an integrated process is recommended, this really means an overarching process with multiple subprocesses. A single organization should be responsible for each subprocess. Different subprocesses may have different organizational leads. A candidate overarching process is described in detail in the section below entitled "Proposed Air Force Intelligence, Surveillance, and Reconnaissance Capability Planning and Analysis Process." An example of a subprocess is the materiel Solution Analysis process led by the Air Force Materiel Command (AFMC). Lastly, in many but not all cases the ISR CP&A process will feed the Air Force Corporate Process, in which components of the ISR process will be assigned to panels (e.g., ISR communications to the Communications Panel). During the Air Force Corporate Process, the impact of board and panel decisions on the overall required ISR capability should be continuously monitored to preclude, for example, a panel's failing to fund a key ISR capability component that may have a low priority as far as that panel is concerned but is vital to the fulfillment of a high-priority ISR need (e.g., a communications link). This can best be self-monitored by establishing a set of interface or giver/receiver relationships across the Air Force. These interfaces become part of a set of agreements on what another board or element is expected and committed to perform. Updating these agreements on a regular basis, with signature concurrence from both sides of the interface, allows timely responses and should result in establishing areas of higher risk when the risk is not the usual technical or schedule risk but rather can be expressed in terms of the risk of an activity's being funded. Mitigation plans for these risks may be developed in much the same manner as for technical or schedule risk.

Recommendation 4-3. The Air Force should adopt the proposed ISR CP&A process by incrementally building on its existing process using pilot projects. The scope of each pilot project should be compatible with available resources, be relevant to both current and future mission scenarios, and include metrics to measure achievement of the desired improvements (e.g., manpower reductions and increased timeliness).[2]

Rationale: The committee was sensitive to the investment that the Air Force has made in its current ISR CP&A process and to the resource-constrained environment in which the Air Force currently finds itself. Hence, the committee designed the proposed process in a manner that would allow the Air Force to expand the capabilities of the current process incrementally over time in response to real-world demands and resource availability. The Air Force is encouraged to develop incrementally, as needed, a reusable information repository and associated suite of analytical tools, models, and simulations that can be used to automate the exploration of trade space of various ISR architectures against mission requirements and cost profiles. This capability would support the sharing of ISR capability information and metrics across multiple analyses and should be retained and evolved from one planning cycle to the next. The Air Force would thus develop an institutional knowledge base of ISR capability and analysis that would allow more rapid and effective decision making. The ISR Capabilities Analysis Requirements Tool (ISR-CART) is a substantive start for this reusable information repository; the Air Force should build on it by populating various tools, models, and simulations that would execute using the shared information. Funding for the pilot project(s) would most likely come from the Air Force organization responsible for the ISR CP&A process. Lastly, a pilot project would provide insight into how challenging the recommended process improvements would be to implement throughout the entire ISR CP&A process.

DESIRED ATTRIBUTES OF AN ENTERPRISE-WIDE INTELLIGENCE, SURVEILLANCE, AND RECONNAISSANCE CAPABILITY PLANNING AND ANALYSIS PROCESS

As described in Chapter 2, the current Air Force ISR CP&A process has several strengths. It is based on a concerted effort by the Air Force to organize a comprehensive planning approach for the ISR portfolio—a process that began with the ISR Flight Plan, evolved into the ISR CP&A process, and broadened to include the Global Integrated ISR (GIISR) Core Function Lead Integrator (CFLI) process. The current processes endeavor to be inclusive and collaborative by coordinating

[2]The proposed process is described in this chapter in the section entitled "Proposed Air Force Intelligence, Surveillance, and Reconnaissance Capability Planning and Analysis Process." Also, notional scenarios are discussed in Chapter 1; they range from regional conflicts (Persian Gulf and Pacific Rim) to global, non-traditional conflicts, to homeland security scenarios.

inputs across relevant CFLIs and associated Major Commands (MAJCOMs). The process also maintains a fairly comprehensive repository of ISR needs, capabilities, and gaps in the ISR-CART tool.

The current processes tend to focus somewhat narrowly on needs that require "big iron" air platform solutions rather than on capability-based needs that might be better addressed by non-airborne solutions.[3] Not surprisingly, the current process also focuses on solutions to near-term problems associated with the ongoing counterinsurgency (COIN) conflict in Southwest Asia. In addition, the process appears to lack a systematic methodology for ensuring that the needs, gaps, and solutions of non-Air Force organizations—particularly those of the intelligence community (IC)—are factored into the mix. In short, the current process does not address all domains, all relevant operational scenarios, or all ISR customers and providers. These and other shortfalls, summarized at the conclusion of Chapter 2, combined with the many strengths of the current process and the best practices of government and industry, lead the committee to articulate the following set of attributes that it believes should serve as guiding design criteria for an enhanced ISR CP&A process.

In order to meet the range of ISR demands described in the Chapter 1, the ISR CP&A process should take advantage of all potential sources, including the following: the IC (e.g., the National Reconnaissance Office [NRO]), non-traditional sources (e.g., non-ISR imaging satellites, F-22, F-35), the Joint community, and coalition partners. It should also address the air, space, and cyberspace domains to fulfill the requirements of the diverse community of users who both depend on and contribute to Air Force ISR. It should include the operational, tactical, and strategic levels of war to provide ISR support for the wide range of Combatant Command (COCOM) Operation Plans and current operations. And it should consider all assets in the sensor-to-user chain, including communications links and the downstream planning and direction, collection, processing and exploitation, analysis and production, and dissemination (PCPAD) portions of the chain.

The ISR CP&A process should be responsive and credible. A responsive process should support a span of requests, from quick-look analyses to deliberative, longer-term analyses that support the Future Years Defense Program (FYDP). To be credible, the process should be transparent, the outcomes should include divestitures as well as additions, and the results should be traceable and repeatable. Table 4-1 summarizes the above attributes. Following is a detailed description of a proposed process that the committee believes will embody these attributes.

---

[3]Col Brian Johnson, Chief, ISR Plans and Integration Division (AF/A2DP), Headquarters, U.S. Air Force. "Air Force ISR CP&A Overview." Presentation to the committee, October 6, 2011.

TABLE 4-1 Desired Attributes of a Comprehensive Intelligence, Surveillance, and Reconnaissance (ISR) Capability Planning and Analysis (CP&A) Process

Desired ISR CP&A Process Attribute	Description
Encompasses all ISR missions	While the process may be primarily driven by near-term mission needs such as the current fight in Southwest Asia and by the counterinsurgency battle doctrine, it should also be able to address future conflicts informed by new doctrines such as Air Sea Battle. The process should look beyond theatre-specific Air Force missions to incorporate those of Joint Forces and the IC around the globe as well.
Addresses all ISR domains and sources	The Air Force should broaden the process aperture to address all domains (including air, space, cyberspace). The process should also address the complete range of multi-INT data sources provided by both traditional and non-traditional platforms.
Includes all ISR assets in the sensor-to-user chain	All end-to-end capabilities required to produce quality intelligence rapidly for the warfighter, including sensors, platforms, data links, and planning and direction, collection, processing and exploitation, analysis and production, and dissemination (PCPAD), are addressed holistically.
Collaborates with ISR-related entities	Key stakeholders and/or representatives in ISR mission and/or scenarios need to be involved in the analysis process in order to take advantage of synergies and to ensure that Joint ISR mission needs are met.
Provides traceability from process inputs to outputs	The underlying data, assumptions, and models used to generate outputs at each stage of the analysis process are revealed.
Is mission/ scenario-based	Mission needs and operational constructs are explicitly factored in to the gap analysis and prioritization.
Is repeatable and enduring	The process is sufficiently simple and transparent to ensure that the Air Force can repeat it on both short and long timescales. The process endures through inevitable changes in Air Force leadership, organization, strategy, and budgets.
Supports trade-off analyses	The decision maker is allowed to rigorously trade off the costs, risks, and utility of alternative ISR force mixes.
Is scalable in size, time, and resolution	A "multi-resolution" ability is provided in order to quickly answer investment questions at a coarse level of resolution, or more deliberately analyze answers at finer resolution. "What-if" analyses are supported for a quick response to focused, specific investment questions as well as large-scale scenario-based investment questions for budget deliberations.
Reduces labor and cost over time	Automated tools are leveraged to produce faster, more accurate results with fewer resources.

PROPOSED AIR FORCE INTELLIGENCE, SURVEILLANCE, AND RECONNAISSANCE CAPABILITY PLANNING AND ANALYSIS PROCESS

This section presents and describes major elements of a recommended, enterprise-wide ISR CP&A process. Suggested methodologies and time lines for reviewing and assessing information are briefly discussed for each function. Tools that may help automate and accelerate the process are offered. As a reminder, the proposed process will provide actionable ways to improve the ISR CP&A process to yield an effective end-to-end investment approach for the integrated mix of ISR capabilities across all domains.[4]

A graphical representation of these vectors in a proposed process diagram is provided in Figure 4-1. Additions to the current process are shaded in red. The two "big vector" changes recommended to the current process are these: (1) the inclusion of a Problem Definition and Approach function that sets up the planning and analysis process, and a (2) Multi-resolution Gap Analysis function that provides the ability to rigorously trade off the costs, risks, and utility of alternative ISR force mixes in an end-to-end system context. The major functions depicted in Figure 4-1 are described in more detail in the following sections.

Problem Definition and Approach

As described in Chapter 2, the current ISR CP&A process is designed to consider *all* needs and gaps on a periodic basis. It is not designed to rapidly answer "what if" questions posed by decision makers faced with urgent issues. A good example of the need for "quick-turnaround" analysis capabilities recently occurred when the chairman of the House Armed Services Committee, in a letter to the Secretary of Defense, expressed the committee's concern that the proposal of the Department of Defense (DoD) to stop purchasing Global Block 30 aircraft was "entirely budget driven with no underlying ISR analysis to support the U-2's ability to fill the gap."[5] The current process, which is designed to consider *all* needs and gaps on a periodic basis, does not have the "machinery" to address individual questions rapidly.

In order to provide the ability to address specific issues, the committee recommends initiating the process with a PDA step—shown in Figure 4-2. The primary

[4]It would be counterproductive to attempt to recommend a highly detailed ISR CP&A process here. The committee does not possess the in-depth knowledge that would be needed regarding all the required process attributes for it to be able to offer effective, actionable suggestions for detailed elements of a comprehensive process. Instead, the committee offers a high-level view that describes "big vectors" that any effective, enterprise-wide ISR CP&A process should possess.

[5]Letter from the Chairman of the House Armed Services Committee to the Secretary of Defense, May 11, 2012.

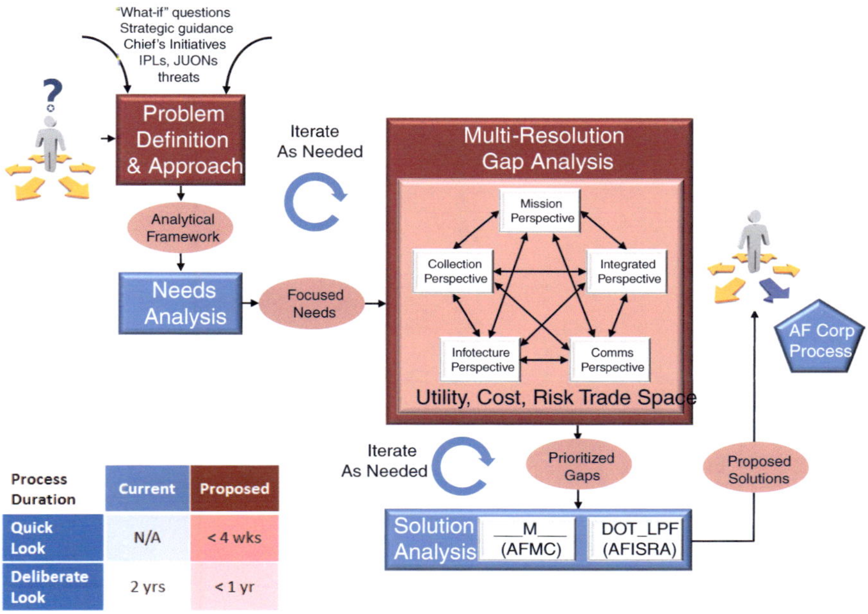

FIGURE 4-1 A high-level diagram showing the major elements of the committee's proposed Air Force intelligence, surveillance, and reconnaissance Capability Planning and Analysis process. NOTE: Boxes and ovals shaded in red represent additions or modifications to the current process, depicted in blue. The table (lower left) indicates anticipated time lines for executing the process. The process is not intended to be strictly sequential in nature. Iterations may occur between various process functions as the analysis evolves. Acronyms are defined in the list in the front matter.

product of the PDA step is an analytical framework designed to guide and focus subsequent steps of the proposed process, shown in Figure 4-1. This framework is carefully crafted through knowledge-elicitation sessions with key stakeholders to drive the ISR CP&A process toward answers that enable and justify investment and divestment decisions. The analytical framework captures, among other things, decision makers' questions, relevant documents, metrics, scenarios, models, and analysis tools needed to configure and support the downstream Needs Analysis and MGA steps in the ISR CP&A process.

In the proposed PDA step, investment questions would be carefully developed and documented through literature research and knowledge-elicitation sessions with decision makers and relevant stakeholders. Literature research includes an analysis of relevant strategic guidance, such as National Intelligence Estimates, the FYDP, Air Force Chief of Staff Initiatives, and Global Threat Analyses. The resulting set of focused questions would guide collaboration among analysts and various domain experts, such as financial analysts, engineers, operators, and intel-

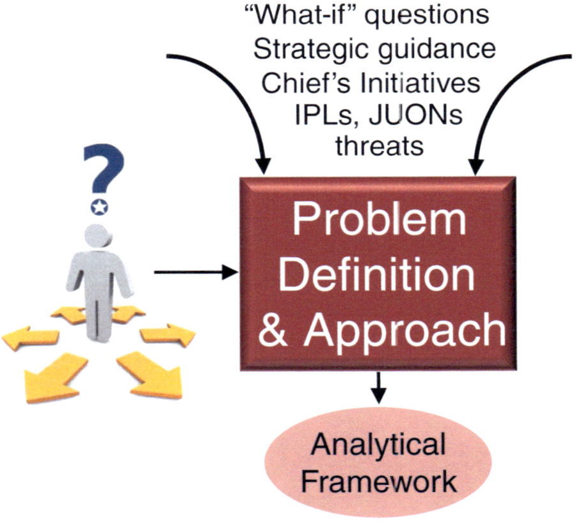

FIGURE 4-2 The Problem Definition and Approach step in the proposed Capability Planning and Analysis process. (See the high-level view of the process in Figure 4-1.)

ligence analysts during Multi-resolution Gap Analysis. If the questions are properly designed up-front, the downstream process should yield quantitative answers that enable decisions in short or long time intervals, with less or more confidence depending on the selected resolution of the analysis.

For example, it is likely that Air Force ISR decision makers have received strategic guidance causing them to consider investment decisions about the ISR capabilities needed to support jungle operations in areas that prohibit overflight. In preparation for assessing the ISR capabilities needed in such contexts, the Problem Definition and Approach team might review various mission needs documents indicating that high-priority jungle operations monitor the movements and actions of guerilla warfare factions under triple-canopy foliage in order to understand their methods and procedures.[6] The team would then work with various stakeholders to elicit specific questions that guide the analysis process toward answers needed to support investment decisions. In the simple example above, one might imagine framing questions such as this: Do we have a standoff sensor in the inventory that can penetrate jungle foliage and detect vehicles? Assuming that there is such a standoff sensor in the inventory, these questions might follow: Do we have communication and data links of sufficient bandwidth in place to support command-

[6]National Research Council. 2009. *Sensing and Supporting Communications Capabilities for Special Operations Forces*. Washington, D.C.: The National Academies Press.

and-control and data exfiltration needs? Do we have a PCPAD capability for the foliage-penetrating sensor data? How well do these solutions work? What are the costs and risks associated with acquiring and/or re-deploying known platform, sensor, communications, and PCPAD solutions?

Once a set of specific questions is developed, the PDA team would construct an analytical framework to focus subsequent steps in the process on answers to these questions. The framework would embody the mission needs in an operational scenario built from studying documents like existing operations plans and foliage maps for the area of interest. The analytical framework would describe the desired performance factors, such as payload standoff distances, sensor resolutions, and communications and data link bandwidth and latencies. The framework would also likely include a set of engineering models configured for the desired operational scenario. The model set might include physical models for propagation at various frequencies; communications models for communications alternatives, such as mobile ad hoc network, remotely piloted aircraft (RPA), and satellite communications; and foliage-penetrating (or poking) Moving Target Indicator (MTI) capabilities for tracking vehicles and dismounts and vehicles under dense foliage. In as many cases as possible, models would be chosen for their ability to scale, to support both coarse-resolution, quick-look analyses and finer-resolution, deliberate-look analyses.

Performance measures and automated tools are also included in the analytical framework. Performance measures (e.g., "average track length" for an MTI sensor) are chosen to help assess the cost, risk, and utility of payload, sensing, communications, and PCPAD alternatives. Automated or semi-automated tools are chosen for their ability to use the models, operational scenario, and performance measures. One example of such a tool is the physics-based Monte Carlo simulation tool used by RadiantBlue to assess the efficacy of alternative platforms, sensing modalities, and communications solutions.[7]

The initial set of models available to support "multi-resolution" may be somewhat small compared to the broad range of investment questions that decision makers might ask. Initially, models might need to be supplanted with subject-matter experts. Over time, it is envisioned that there would be built a library of reusable models that can be stored in an ISR Analytic Information Repository from which they can be easily retrieved and reconfigured to support a wide range of analyses.

With the above example in mind, the Problem Definition and Approach phase can be completed in few days for well-understood issues or for quick-look analyses in which coarse answers are needed quickly to support, for example, an urgent planning, programming, or budgeting question. Alternatively, a deliberate deci-

[7]Larry Shand, President, RadiantBlue, Inc. "RadiantBlue Modeling and Simulation Capability." Presentation to the committee, January 5, 2012.

sion analysis phase can take weeks filled with workshops designed to interactively elicit information and decision needs from government stakeholders or staffs. For either quick-look or deliberate decision analysis approaches, this phase of the process should allow time for the performance of a limited trade-off analysis based on existing information and input from available subject-matter experts in order to produce coarse, initial answers to questions that might be sufficient for decision makers' needs, and to develop a more informative analytical framework for subsequent phases of the process.[8]

The PDA step should be led by experts with experience eliciting actionable questions and related information from stakeholders in a form suitable for downstream analysis. The size and composition of the expert team will likely vary over time in response to changing investment decision needs. The team will likely consist of a small number of permanent members who have access to government, industry, and academic experts available on call to provide assistance on an as-needed basis. The organization responsibility for maintaining the team, funding requirements, and other team characteristics should be considered during the pilot project phase of the process implementation.

Needs Analysis

The primary purpose of the Needs Analysis step, shown in Figure 4-3, is to transform the set of investment questions contained in the analytic framework into focused needs that can be rigorously analyzed during Multi-resolution Gap Analysis. Unlike the Needs Analysis step in the current ISR CP&A process, which produces an unconstrained list of needs, the committee's proposed Needs Analysis step would produce a constrained list of capability needs designed to focus the process on answers to specific questions.

Needs Analysis for a deliberative, long-term capability-planning exercise might employ subject-matter experts to assess likely scenarios provided by the analytical framework and make recommendations about needs for these scenarios. In the case of the earlier "triple canopy" example, needs might focus on the reconnaissance of a group of guerilla fighters, or on the exfiltration of unattended ground-sensor (UGS) data. More specifically, the Needs Analysis step might express the previous question—Do we have a standoff sensor in the inventory that can penetrate jungle foliage and detect vehicles?—in the form of the specific, analyzable need: We need the ability to detect 5-meter-square objects under triple-canopy jungle foliage at a range of 20 km. Although specificity is helpful, the Needs Analysis team should

[8] Doug Owens, Manager, Enterprise Analysis, Defense Business Unit, TASC. "An Enterprise Approach to Capability-Based Analysis: Best Practices, Tools, and Results." Presentation to the committee, January 5, 2012.

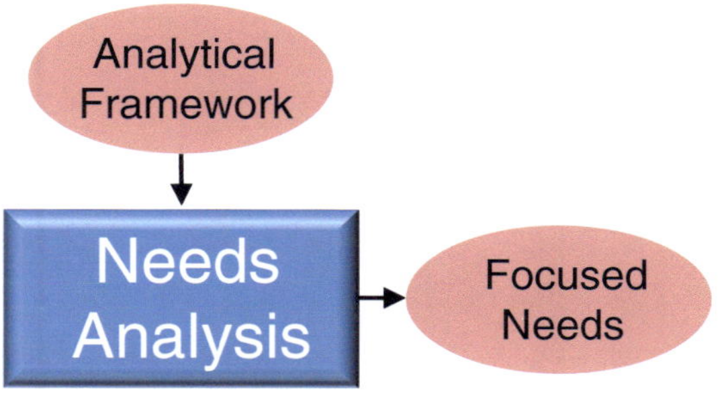

FIGURE 4-3 The Needs Analysis step in the proposed Capability Planning and Analysis process. (See the high-level view of the process in Figure 4-1.)

avoid the temptation to express the need as a solution. For example, it would not be appropriate to express the same need exemplified above as: The Air Force needs a gimbal-mounted Geiger-mode LIDAR mounted on a Predator remotely piloted aircraft that can detect 5-meter-square objects under double-canopy jungle foliage at a range of 20 km. Disguising solutions as needs would minimize the value of the trade-space machinery used in the next step of the ISR CP&A process.

A focused needs list reduces process time and cost and provides the ability to answer specific questions rapidly. That said, the process should also periodically assess a wider set of needs so as to prevent the Air Force from myopically focusing on the mission of the day. The Air Force should consider reviewing all needs on a 3-year cycle, with no attempt to prioritize or filter needs during these broader looks.[9] By avoiding the temptation to eliminate seemingly extraneous or unimportant needs during the process, the Air Force would retain an inventory of stated needs that might become relevant under a different set of strategic assumptions or operational scenario conditions. This would also help maintain a holistic view, which ensures that the perspectives of non-Air Force organizations do not wind up "on the cutting-room floor."

The primary participants in the Needs Analysis function are the COCOMs, which typically express their needs by means of Integrated Priority Lists; the MAJCOMs, which represent the interests of their affiliated COCOMs; and the national IC. Because the IC, rather than the Air Force Space and Missile Systems Center, provides the majority of space-based ISR capabilities, it is imperative that the Air Force ISR Agency (AFISRA) reach out to the IC so that the ISR needs of their respective stakeholders can be shared in a disciplined, systematic way. Accordingly, the Air Force should actively work with the IC to formally link the ISR CP&A

[9]A similar approach is used by the National Geospatial-Intelligence Agency (NGA).

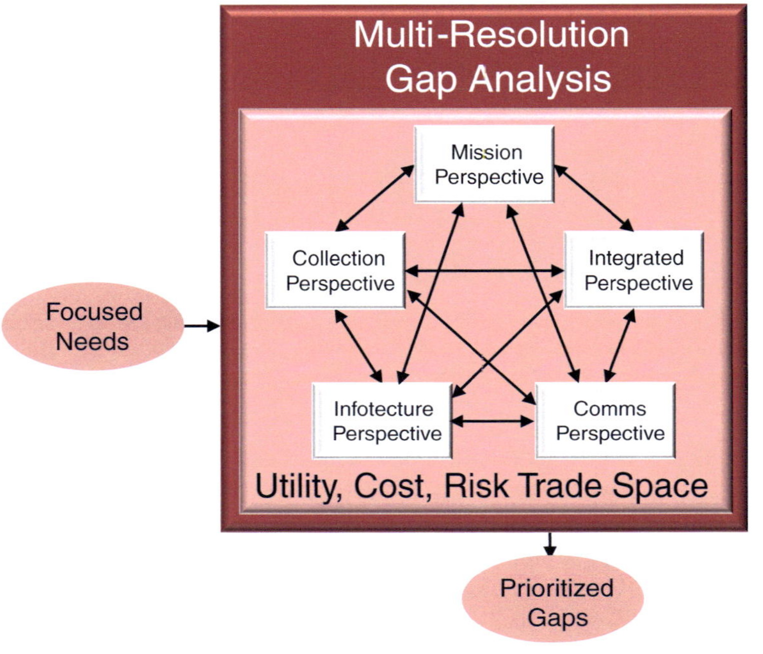

FIGURE 4-4 The Multi-resolution Gap Analysis in the proposed Capability Planning and Analysis process. (See the high-level view of the process in Figure 4-1.)

process with the processes of IC partners, particularly the National Geospatial Intelligence Agency (NGA), the NRO, and the National Security Agency.

Needs should be documented in writing using structured fields that can be stored and easily searched using a database-management system. At a minimum, needs records should include a unique identifier, description, and measures of effectiveness and/or measures of performance that allow the Air Force to quantitatively assess progress toward meeting the desired end state. Finally, as is the case today, the Air Force should continue to capture all needs in the ISR-CART database. Although the process of accurately and completely documenting, gathering, reviewing, and updating needs records is potentially tedious and time-consuming, once the initial set of needs is entered into the ISR-CART database, future Needs Analysis activities might be shortened considerably through the updating of the initial set on a change-only basis.

Multi-resolution Gap Analysis

The primary objective of the Multi-resolution Gap Analysis step, shown in Figure 4-4, is to rigorously compare focused needs with existing capabilities and to produce a list of prioritized gaps in a trade space that assesses mission utility, cost, and risk from multiple perspectives and at multiple levels of fidelity. The MGA step is a highly collaborative process among analysts, domain experts, and related stakeholders whose shared perspectives span a broad ISR capabilities trade space. MGA is a modeling and simulation-based approach to problem analysis that allows the Air Force to examine complex problems systematically by quantitatively analyzing the problem from different perspectives and at different resolutions. The approach uses a wide variety of different models and tools tailored for each perspective. It integrates different capability perspectives in a manner that provides traceability between causes and effects across the ISR enterprise.[10]

Exploring the problem from multiple domain perspectives is valuable because different domains often describe the problem with different representations and semantics, making it difficult to "understand the elephant" when described by just one of its many domain parts. Exploring the problem at multiple resolutions provides the ability to rapidly develop answers to questions using coarse, low-resolution models, or to deliberately produce accurate answers over longer time intervals using fine, high-resolution models. For example, it is possible to obtain a coarse estimate of the volume of water in a lake by multiplying the lake's maximum depth, width, and length. Or, a much more accurate estimate can be produced using a high-resolution model that sums small volumes estimated at each point on a grid laid across the lake floor.

MGA approaches are particularly useful when the high- or low-resolution models or their input data are infused with uncertainty. These circumstances are common in planning for capabilities associated with military applications for which models of complex combat operations do not exist—and even if they did, would likely suffer from highly uncertain inputs associated with the fog of war.[11] Theoretically sound, well-constructed models are consistent across resolution levels. And a well-designed Multi-resolution Gap Analysis approach scales in order to address large, complex problems, albeit often with less detailed and accurate solutions.

MGA is not a model or methodology within a model, but a methodology for the application of multiple models. The capabilities within the MRA methodology would not include models in which resolution could be dialed for applications. Mixing levels of resolution within models is not a desired practice. MGA's driving concept is that high-resolution models can generate metrics that are rolled into

[10]Doug Owens, Manager, Enterprise Analysis, Defense Business Unit, TASC. "An Enterprise Approach to Capability-Based Analysis: Best Practices, Tools, and Results." Presentation to the committee, January 5, 2012.

[11]James H. Bigelow and Paul K. Davis. 2003. "Implications for Model Validation of Multiresolution, Multiperspective Modeling (MRMPM) and Exploratory Analysis." Santa Monica, Calif.: RAND. Available at http://www.rand.org/pubs/monograph_reports/2005/MR1750.pdf. Accessed March 22, 2012.

lower-resolution models. The various levels of resolution come into play in terms of the layering of models that employ more and more aggregation to meet decision needs sufficiently and responsively. Analysis can be conducted at lower levels of resolution for aggregate levels of trade-offs if such information is sufficient to aid decisions at those levels, as long as those aggregate-level effects maintain traceability to higher-resolution performance factors.

The MGA construct may consist of numerous models and tools to flesh out trade-offs sufficiently to support decisions. For example, a series of tools may be appropriate to assessing cyberthreat impacts to missions and determining options for mitigation of high-order effects. Technical, high-resolution nodal analysis of a physical network could determine the pervasiveness of a specified cyber attack. That analysis could produce metrics on estimated network downtime and degree of degradation to specific segments of the network architecture. Rather than carrying the high-resolution effects of the cyber virus to the next level of analysis, only the effects-based metrics would be carried forward. These could then serve as planning factors in a process flow model to determine data-throughput impacts of critical information delayed by the cyber effects on segments of the architecture. Those throughput-metric results could then serve as uncertainty bands in a simulation of the affected missions, such as communications support to joint operations in a specified scenario.

The MGA box in Figure 4-4 shows four domain perspectives and one integrated perspective, all interconnected to emphasize that the MGA process requires joint and several interactions among all perspectives. The four independent perspectives initially suggested when analyzing the utility, cost, and risk trade-off space are the *Mission Perspective,* which focuses on operational plans, force structure, and the command-and-control of operational assets; the *Collection Perspective,* which primarily includes ISR sensors and platforms; the *Infotecture Perspective,* which focuses on PCPAD activities in support of operations; and the *Comms Perspective,* which includes cyber, network, and communications capabilities that enable command and control and sensor data exfiltration. Additional perspectives undoubtedly exist and bear consideration when circumstances require.

More importantly, MGA uses an *Integrated Perspective* that allows trade-offs in utility, cost, and risk among the other perspectives. Tailored and scaled to the needs of the decision maker, the iterative, integrated process provides quick-look assessments through streamlined analysis processes early in the analysis cycle. And it increasingly adds layers of fidelity that allow broader and deeper analyses of the ISR capability trade space. Integrating different domain perspectives at different resolution levels then provides an enterprise view of needs associated with existing or missing capabilities. Examples of methods used to integrate operations and cost perspectives include multi-attribute utility analysis, which integrates multiple metrics into a set of value metrics, and inference analysis, which maps capability performance metrics into a spider web mosaic for diagnostic assessments of potential impacts of variations in each metric.[12]

Many of these methods employ hierarchical structures, in which each layer has its own elements, properties, relations, and metrics that characterize the important behavior or operations within that layer. Elements of the layer may depend on other layers in the hierarchy, such as the interdependencies between communications and ISR. The attributes or metrics of these elements can be mapped through transformations that translate those attribute dependencies onto the next layer's elements or behaviors. The result is a layering of capabilities and metrics that begin at the top in a more aggregated, higher level of abstraction, flowing down to lower (higher-resolution) layers, enabling trade-offs that are still rooted in the detailed physics levels but understood and assessed at higher decision levels. By this transformation layering, attributes from one layer to another can be decoupled and re-characterized to enable analysis at higher levels that, though rooted in the high-resolution physics of technologies and systems, are not strictly linked to specific parameters of individual systems or concepts. As long as the transformations that enable mapping from one layer to another can be constructed, the relations can be preserved and tracked through the entire enterprise.

MGA brings all perspectives to bear in order to find a viable solution, with viability conditioned by utility, cost, and risk assessments. These assessments provide separate, quantitative insights using interactive, model-based analyses driven by capability metrics assigned to PCPAD information and data flows; command, control, communications, and computer network trade-offs; sensors and platforms; operational concepts of operation; and cyber/information operation impacts. They also map capability metrics to cost estimating and risk-analysis trade-offs, and project costs over planning horizons. Cost-benefit analyses can be developed for individual domain perspectives as well as for combined families of systems, architectures, or the entire enterprise. Initially coarse, cost-benefit assessments are refined during later stages of the Multi-resolution Analysis, and subsequent Solution Analysis. A good example of a multi-resolution financial analysis capability is provided by TASC's Financial and Business Analytics tool, which maps capability metrics to cost estimating and risk-analysis trade-offs, plus cost projections over planning horizons (Analysis of Alternatives, Program Objective Memorandum [POM] inputs).

Continuing with the previous jungle operations example, an RPA solution might be preferred on a cost and accuracy basis, but it might present a risk by indicating Air Force presence and interest unless flown at high altitudes to avoid detection. Risks might lie in the trade space between standoff distance, detectabil-

---

[12]TASC has developed an example of a methodology and associated tool—called Integrated Decision Analysis—that enables decision trade-offs among risks, sensitivities, and programmatic considerations across multiple perspectives.

ity by the adversary, and the wait for the development of a less-observable RPA capability. The mission perspective would influence the trade space in multiple ways. For example, high-resolution video (part of the collection perspective) would require "boots-on-the-ground" (meaning that the collection perspective affects the integrated perspective, since a Joint capability, e.g., a UGS, may be needed). Exfiltration of the video would require taking the communication perspective into account—for example, a data network consisting of ground elements connected to airborne and space relays to get the data into the hands of analysts and decision makers. This information flow from the theater to the users reflects the analysis and dissemination represented by the infotecture perspective and also illustrates the end-to-end analysis designed into the process. This process can be (and likely will be) iterated—for example, the use of robotic UGSs may reduce risks to human operators but may demand the development of new Air Force air deployment capabilities and communications architectures.[13]

MGA is executed as an interactive collaborative process among analysts, subject-matter experts, and various process stakeholders. The key to achieving a flexible, robust analytic capability is founded on the application of quantitative, model-based methods that allow an examination of the entire enterprise from different capability perspectives, integrated for a complete view of total capability. The use of model-based analytics within an integrated, interactive process allows one to determine at various times and complexity scales whether the capabilities exist to satisfy the stated needs, or, if gaps exist, what their order of priority is for subsequent consideration in the solutions analysis phase of the process.

An MGA process is envisioned that is jointly funded by and distributed among AF/A2, AFISRA, and Air Combat Command (ACC), and is at various times led by either the AF/A2 or the GIISR CFLI. For example, AF/A2 may choose to lead the process when responding to a "short-fuse request," from the Office of the Secretary of Defense, the Office of the Director of National Intelligence, or the Congress, to develop and justify a budget position. And the CFLI may be better suited to lead periodic, deliberate analyses of a broader set of needs and gaps on a POM planning cycle. The distributed components of the MGA process include the ISR-CART database, a repository of reusable models that can be accessed by means of metadata stored in the ISR-CART, and the addition of a variety of modeling and simulation and other tools needed for assessing the utility, cost, and risk of ISR capabilities within and across domain perspectives that could become the substrate of an ISR CP&A Analytics Repository. Because the ISR-CART is currently maintained

[13]J.M. Smith, M. Olivieri, A. Lackpour, and N. Hinnerschitz. 2009. "RF-mobility Gain: Concept, Measurement Campaign, and Exploitation." *IEEE Wireless Communications* 16(1):38-44. Available at http://repository.upenn.edu/cgi/viewcontent.cgi?article=1435&context=cis_papers. Accessed March 22, 2012.

by AFISRA, it might be easiest in the near term to build and maintain an initial repository and tool set at AFISRA. Over time, however, a distributed, networked capability is envisioned that can be readily accessed through the Internet by all MGA participants and their designated contractors. For example, one might envision a distributed, physics-based modeling-and-simulation capability that might use communications models maintained by a group in Los Angeles, infotecture models from a group located in San Antonio, Texas, and platform and sensor models maintained by a group in Dayton, Ohio.

The core MGA team would consist of "on-call" subject-matter experts in each of the perspective domains, and an "integrator" with the breadth of skills and experience in Multi-resolution Analysis required to lead and manage the process. Because of the potentially large degree of iterative interactions between the PDA and Needs Analysis steps, it may be beneficial to have the integrator also serve as the PDA lead. Doing so might shorten the lines of communication so as to speed the process, reduce confusion, and manage costs efficiently. The domain experts would most likely not be co-located. And networked elements of the integration team might also collaborate from multiple locations, such as Washington, D.C.; Langley, Virginia; and San Antonio, Texas.

With the many variables associated with implementing an MGA process, it is unclear how much to suggest that the Air Force annually budget in order to fund the MGA team or the purchase and maintenance of models, tools, and so on. It might be best to begin with a small "calibration" project, co-funded by AF/A2 and the GIISR CFLI, that would help them gain an understanding of the major cost drivers and use lessons learned to refine future MGA requirements. Working together, the two organizations would develop a plan and budget for expanding and improving the process over time.

Solution Analysis

The primary purpose of the Solution Analysis step of the ISR CP&A process, shown in Figure 4-5, is to analyze and recommend materiel and non-materiel solutions that fill prioritized gaps provided by the MGA phase. As shown in Figure 4-5, the Air Force Materiel Command is generally responsible for developing materiel solutions, whereas AFISRA develops non-materiel solutions. Recommended solutions are forwarded to decision makers who work within the Air Force Corporate process to prioritize and seek funds to implement them.

It is important to note that both the AFMC and the AFISRA rely on established processes for assessing and developing solutions. For example, AFMC's Development Planning community has an established capability planning and assessment process that analyzes solutions, with capability management teams consisting of stakeholders across the science and technology, acquisition, and operational com-

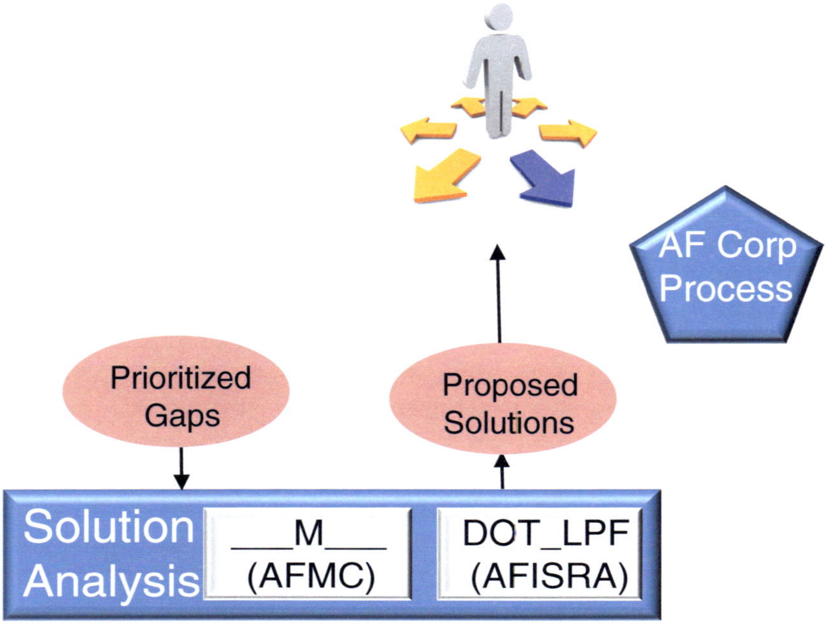

FIGURE 4-5 The Solution Analysis step in the proposed Capability Planning and Analysis process. (See the high-level view of the process in Figure 4-1.)

munities. Rather than calling for a reinvention of the process wheel, the committee recommends that the Air Force continue to rely on established AFMC and AFISRA processes to assess and develop solutions for capability gaps. That said, there is a need for a decision-making organization to determine initially whether a particular gap requires a materiel or a non-materiel solution. In addition, such an organization should assign MAJCOM owners to gaps and analysis solutions, serve as an interface and advocate with Air Staff and other stakeholders when required, and generally oversee the analysis of the various materiel and non-materiel solution processes. Because AF/A2 would primarily invoke the ISR CP&A process to seek rapid answers to "what if" questions that do not require deep solutions analyses, the GIISR CFLI might be the best organization to oversee what can sometimes be protracted materiel and non-materiel solution analyses in support of the development of new and significant capabilities.

TABLE 4-2 The Proposed Process Satisfies the Desired Attributes of an Enterprise-wide ISR Capability, Planning and Analysis (CP&A) Process

Desired Process Attribute	Proposed Process Capability
Encompasses All ISR Domains and Entities	
Encompasses all ISR missions Addresses all ISR domains and sources Includes all ISR assets in sensor-to-user chain Collaborates with ISR-related entities	• The Problem Definition and Approach (PDA) step brings the right players to the table to ensure that the process is configured to address all relevant investment questions. • The Multi-resolution Gap Analysis (MGA) framework explicitly incorporates and integrates the perspectives of all relevant domains, sources of ISR needs and capabilities. • Sensor-to-user assets, including platforms, communications and command-and-control links, and PCPAD capabilities are modeled and analyzed by means of simulations to answer investment questions in a performance versus cost versus risk trade space.
Produces Credible Outcomes	
Provides traceability from process inputs to outputs Is mission/scenario-based Is repeatable and enduring Supports trade-off analyses	• The multi-resolution framework ties the analysis process to authoritative data sources and the analysis guidance established during the PDA step. It is executed as an interactive, collaborative process among various analysis elements and players to provide a more complete cost, risk, and utility trade-space analysis. • Complex domains of capability are analyzed from different perspectives with tailored models and tools appropriate for each perspective, but with the various segments of the analysis integrated to provide traceability of cause and effect for combined total impact. • High-fidelity technical performance measures on sensor/system effects are mapped against specific mission scenarios to assess operational impacts and prioritize gaps.
Efficiently Uses Limited Resources	
Is scalable in size, time, and resolution Reduces labor and cost over time	• The multi-resolution framework provides quick-look assessments of capability and cost and/or affordability through streamlined applications of cost, risk, and utility trade-space analyses. The framework adds more and more layers of fidelity where needed to refine quick-look assessments over time. • Automated analysis tools, such as modeling and simulation, reduce the number of people and amount of time required to conduct trade-space analyses. Models are refined, stored, and reused to the greatest extent possible to minimize duplication of modeling effort.

NOTE: Acronyms are defined in the list in the front matter.

CONCLUDING THOUGHTS

The recommended process described in this chapter is intended to enhance, not to replace, the current ISR CP&A process in a manner that achieves the desired attributes listed in Table 4-1. In particular, the addition of a PDA step would allow decision makers to focus the process on specific investment questions while also supporting the periodic need to assess all ISR needs and gaps. Also, the current gap analysis step would be expanded to integrate explicitly the perspectives and resources of multiple domains in a mission context. The proposed gap analysis approach incorporates a multi-resolution framework that allows the process to scale in a consistent manner, from quick-look assessments designed to address urgent questions rapidly, albeit with less fidelity, to deliberate-look assessments that produce higher-fidelity answers at the cost of additional time and resources. Table 4-2 summarizes the proposed process enhancements that satisfy the desired attributes, summarized in Table 4-1, in a robust, comprehensive Air Force intelligence, surveillance, and reconnaissance Capability Planning and Analysis process.

Despite the attempt to design a process that enhances rather than replaces the current process, the addition of the PDA and MGA functions would require careful planning prior to their implementation. Further, the Air Force may wish to implement the process changes in a staged fashion designed to minimize disruptions to the ongoing process. The Air Force is urged to roll out the recommended process enhancements by way of a pilot project, or a series of pilot projects, to lay the foundation of a future process that the Air Force can thoughtfully build on over time to achieve the desired end state.

In summary, the value inherent in achieving this end state derives from its ability to effect the following: (1) enhance the quality, transparency, repeatability, and credibility of proposed investments; (2) provide greater insight into cost, risk, and mission utility assessments; (3) scale from quick-look through long-term analyses; (4) expand consideration and analysis of Joint and interagency capabilities; (5) more fully address all ISR domains (air, space, land, maritime, cyberspace); (6) encompass complete "sensor-to-user" chain including PCPAD; and (7) reduce the time and labor required to answer investment questions.

Appendixes

Appendix A

Biographical Sketches of Committee Members

Brian A. Arnold, *Co-Chair,* is the vice president of Space Strategy for Raytheon Company's Space and Airborne Systems (SAS). In this role, he determines evolving customer needs in the defense, intelligence, and civil arenas and develops strategies to meet them with space-qualified solutions. He also leads planning efforts for expanding core SAS space markets and technologies. Before assuming his current position, he served as the vice president and general manager of Space Systems at Raytheon SAS. A retired U.S. Air Force lieutenant general, he has 35 years of experience in leading space superiority programs and possesses exceptional space market knowledge and expertise. Prior to joining Raytheon in 2005, Mr. Arnold served as commander of the Space and Missile Systems Center, Air Force Space Command, Los Angeles Air Force Base, the nation's center of excellence for military space acquisition. There, he managed the research, design, development, acquisition, and sustainment of space launch and command-and-control systems, missile systems, and satellite systems. Mr. Arnold was commissioned through Officer Training School at Lackland Air Force Base, Texas, in 1971, and spent the majority of his Air Force career in operations as a pilot in FB-111 and B-52 aircraft; he has served as a commander at the flight, squadron, wing, and subunified level of command. As the director of Space and Nuclear Deterrence for the Assistant Secretary of the Air Force for Acquisition, he was responsible for space and missile systems. Mr. Arnold received a bachelor's degree in education from California State University, Hayward, and a master's degree in administrative education from Pepperdine University, Los Angeles.

Lawrence J. Delaney, *Co-Chair,* is currently a private consultant. He retired as the executive vice president of operations and president of the Advanced Systems Development Sector of Titan Corporation. Previously he held distinguished positions with Arete Associates, Inc.; Delaney Group, Inc.; BDM Europe; and the Environmental and Management Systems Group at IABG. He was also the Acting Secretary of the Air Force and served as the Assistant Secretary of the Air Force for Acquisition, as well as the Air Force's service acquisition executive, responsible for all Air Force research, development, and acquisition activities. He provided direction, guidance, and supervision of all matters pertaining to the formulation, review, approval, and execution of acquisition plans, policies, and programs. Dr. Delaney has more than 41 years of international experience in high-technology program acquisition, management, and engineering, focusing on space and missile systems, information systems, propulsion systems, and environmental technology. He served as a member of the National Research Council's (NRC's) Board on Army Science and Technology and chaired the NRC's Air Force Studies Board. He is currently vice chair of the Army Science Board.

Collin A. Agee is the senior advisor for intelligence community engagement in the Office of the U.S. Army Deputy Chief of Staff for Intelligence (G-2). Previously he served for 2 years at the National Geospatial-Intelligence Agency (NGA) as the agency's senior official for future systems, after joining NGA as the deputy director for Future Warfare Systems in January 2009. He was previously the Army G-2's senior intelligence, surveillance, and reconnaissance (ISR) advisor. In 2008, he deployed to Baghdad as Deputy CJ2 Forward, the senior military intelligence officer in the U.S. Embassy, where his duties included providing intelligence to the Multinational Force-I staff, as well as conducting strategic intelligence engagement with the Iraqi intelligence services and senior Iraqi government officials. He previously served as the Army's director for ISR integration for 2 years, following 4 years as a contractor for Booz Allen Hamilton in support of Army G-2 as a member of Task Force Actionable Intelligence in the ISR Integration Directorate and the Army Intelligence Master Plan, where he was a co-author of the Army Intelligence Transformation Campaign Plan. His tenure was highlighted by the conceptualization and implementation of the Actionable Intelligence Initiative, also known as Focus Area 16. He commanded Headquarters Company in the 125th Military Intelligence Battalion, Schofield Barracks, Hawaii, and the Los Angeles Recruiting Battalion. Mr. Agee has a master's in military arts and science from the School for Advanced Military Studies and is a graduate of the Defense Intelligence Agency's Postgraduate Intelligence Program. He is a member of the Armed Forces Communications and Electronics Association's Intelligence Committee and the Executive Committee for the Military Intelligence Corps Association.

Melani Austin is a senior program manager in Advanced Development Programs (ADP) at Lockheed Martin Aeronautics Company, with 29 years of experience in technology development and program execution on low observable (LO) aircraft. She is currently responsible for the design, development, manufacturing, and flight testing of intelligence, surveillance, and reconnaissance (ISR) production programs. Ms. Austin began her ADP career in 1983 as an LO engineer and has continued to be a pioneer in the development of Stealth/Signature Management technologies and their integration into survivable weapons systems, most conspicuously on the F-117A, F-22, and TIER III programs. Throughout her career, Ms. Austin's engineering and program execution efforts have been focused on the development and application of flight-quality LO material technologies and ISR subsystem integration. She graduated with a B.A. in economics from the University of California at Los Angeles.

Thomas J. Burns is the manager of Science Applications International Corporation's (SAIC's) Sensors and Phenomenology Operation, where he is responsible for more than 750 employees and more than $250 million in research and development, system solutions, and products business. Prior to joining SAIC, Dr. Burns co-founded and served as chief executive officer and chair of SET Corporation, a small high-tech business specializing in the creation and commercialization of smart sensing technologies. Under his leadership SET grew, without external investment, to 100 employees and $30 million in annual revenue. Acquired in January 2010, SET operates as a wholly owned subsidiary of SAIC. Prior to founding SET, Dr. Burns co-founded and served as chief operating officer of ObjectVideo, Inc., a venture-capital-backed leader in smart video solutions for commercial and military security applications. He joined ObjectVideo from the Defense Advanced Research Projects Agency, where he pioneered the development of model-based signal and image exploitation technologies, building on his experiences directing Computer Vision research as a U.S. Air Force officer at the Air Force Research Laboratory (AFRL). While assigned to AFRL, he led AFRL's premiere Automatic Target Recognition program, receiving AFRL's prestigious Peter R. Murray Program Manager of the Year Award. Dr. Burns is a co-inventor of patents on video and radar technology and has published numerous refereed papers in areas as diverse as electro-optics and wavelet mathematics. He was a member of the National Research Council's Committee on Sensing and Communications Capabilities for Special Operations Forces and is currently a member the Air Force Studies Board. Dr. Burns also serves as a board director of Yakabod, Inc., an innovative knowledge-management product company, and he serves as a member of the Washington, D.C., Capital Executive Board. He received a Ph.D. in electrical engineering from the Air Force Institute of Technology.

Pamela A. Drew is currently the senior vice president for Strategic Capabilities and Technology at TASC, where she is responsible for corporate strategy, capability development, and business generation in key domains such as cyber and systems engineering and integration. Dr. Drew was previously the vice president of Business Development and Strategic Initiatives, Mission Systems Sector, for Northrop Grumman. Before that, she had been vice president and general manager in various positions of Boeing's defense business, including the Command, Control, Communications, Intelligence, Surveillance and Reconnaissance (C3ISR) Solutions organization, which included the airborne ISR programs such as the Airborne Warning and Control System family, and transformational communications programs. She also spent several years in various positions in Boeing's Phantom Works leading technology strategy and research and development primarily focused on network-centric capabilities. Prior to her years at Boeing, Dr. Drew was an assistant professor in the Department of Computer Science at the Hong Kong University of Science and Technology, where she was part of the founding faculty. Earlier she had worked at US WEST Advanced Technologies as a member of the technical staff, leading database and software engineering research projects. She earned a B.S in mathematics and computer science (1985), an M.S. in computer science (1987), and a Ph.D. in computer science (1991) from the University of Colorado, Boulder, where she was named a Distinguished Engineering Alumni in 2007. She is currently serving as the vice chair of the National Research Council's Air Force Studies Board; as vice chair of the Engineering Advisory Committee at the University of Colorado, Boulder; and on the board of directors at the University of Washington Applied Physics Laboratory. She is the author of more than 30 peer-reviewed technical publications and a featured speaker on network-centric technologies in national and homeland security.

Rand H. Fisher is the senior vice president of Systems, Planning, Engineering, and Quality for the Aerospace Corporation. Prior to that, Rear Admiral Fisher was the vice president and senior advisor, Situational Awareness, for Lockheed Martin Information Systems and Global Services. Prior to his retirement from the U.S. Navy, he served concurrently as director, Communications Acquisition and Operations Directorate within the National Reconnaissance Office; commander of the Space and Naval Warfare Systems Command (SPAWAR) Space Field Activity; Naval Program Executive Officer for Space Systems; and director, Transformational Communications Office. He previously served as commander of the Naval Air Warfare Center Weapons Division in China Lake, California, and as assistant commander for Test and Evaluation at the Naval Air Systems Command. At SPAWAR, RADM Fisher served as program manager for research and development, director of the Systems Program Management Division, lead systems engineer for the Naval Space Technology Program, deputy program manager for the Special Systems Program

Office, major program manager of the Special Systems Program Office, and major program manager of the Advanced Systems Program Office. RADM Fisher graduated from the Naval Postgraduate School with an M.S. in physics. He has been awarded the Distinguished Service Medal, Defense Superior Service Medal, Legion of Merit, Defense Meritorious Service Medal, and various other service medals and awards. He is a former member of the Air Force Studies Board.

Keith R. Hall is a senior executive advisor for Booz Allen Hamilton, having retired as a senior vice president from the corporation in December 2009. He joined Booz Allen in 2002 following a distinguished career in the federal government. From February 1996 to December 2001, he served as director of the National Reconnaissance Office (NRO). In March 1997, he was also appointed by the President and confirmed by the Senate as Assistant Secretary of the Air Force for Space. As NRO director, he was responsible for the acquisition and operation of all U.S. space-based reconnaissance and intelligence systems. Within the Air Force, he was responsible for the overall supervision of space matters, with primary emphasis on policy, strategy, and planning. Mr. Hall has worked in various capacities in U.S. intelligence since 1970, when he received his commission as an officer in the U.S. Army. He served 9 years in Army intelligence, including two tours during which he commanded overseas operational intelligence units. He left the Army in 1979 after being selected a Presidential Management Intern and appointed to the Office of Management and Budget, where he served as budget examiner for the Central Intelligence Agency. In 1983, Senator Barry Goldwater appointed him a member of the professional staff of the U.S. Senate Select Committee on Intelligence, on which he served until 1991. He had primary responsibility for supporting the committee in the annual intelligence budget authorization process and, as deputy staff director, supported all committee oversight activities as well as the formulation of intelligence-related legislation. From 1991 to 1995, Mr. Hall served as Deputy Assistant Secretary of Defense for Intelligence and Security in the Office of the Secretary of Defense. In addition to his responsibilities for policy development, resource management, counterintelligence, and security oversight, he was chair of the National Counterintelligence Policy Board and co-chair of the Intelligence Systems Board. Prior to his presidential appointment, he served as executive director for Intelligence Community Affairs and director of the Community Management Staff from May 1995 to February 1996 at the Central Intelligence Agency. In this capacity he had responsibility for overall policy and resource management of national intelligence activities and was the principal architect and co-chair of the Intelligence Program Review process, he co-chaired the Security Policy Forum, and co-chaired the study group that created the National Imagery and Mapping Agency. Mr. Hall earned an M.A. in public administration from Clark University and an honorary doctorate from Alfred University in New York.

Leslie F. Kenne is a private consultant and president of LK Associates. Previously, she held the position of Deputy Chief of Staff for Warfighting Integration, Headquarters, U.S. Air Force. During her military career, she held the positions of commander, Electronic Systems Center, Hanscom Air Force Base, Massachusetts; and program director for the Joint Strike Fighter, Headquarters, U.S. Air Force. Lt Gen Kenne received her M.S. in procurement management from Webster College. She also attended the U.S. Air Force Test Pilot School; the National War College; the Defense Management College at Fort Belvoir, Virginia; the Whittemore School of Business and Economics at the University of New Hampshire; and the John F. Kennedy School of Government at Harvard University.

Robert H. Latiff retired as a major general from the U.S. Air Force and is currently a private consultant and the president of R. Latiff Associates, providing advice on advanced technology matters to corporate and government clients and universities. He also holds an appointment as research professor and director of the Intelligence and Security Research Center at George Mason University. Prior to joining George Mason University, Dr. Latiff was vice president and chief technology officer of Science Applications International Corporation's space and geospatial intelligence business. Dr. Latiff is the chair of the National Research Council's National Materials and Manufacturing Board and a member of the Air Force Studies Board. He has led and participated in numerous studies, and he writes and speaks frequently about critical materials and processes. Dr. Latiff is also an active member of the Intelligence Committee of the Armed Forces Communications and Electronics Association. Major General Latiff's last active-duty assignment was at the National Reconnaissance Office, where he served as deputy director for Systems Engineering and director of Advanced Systems and Technology. He has served as vice commander of the USAF Electronic Systems Center and as commander of the North American Aerospace Defense Command (NORAD) Cheyenne Mountain Operations Center in Colorado Springs, Colorado. General Latiff received his commission from the Army Reserve Officers' Training Corps program at the University of Notre Dame. He entered active service in the U.S. Army and later transferred to the U.S. Air Force. General Latiff has served on the staffs of the Headquarters of the U.S. Air Force and the Secretary of the Air Force. He received his Ph.D. and his M.S. degrees in materials science and his B.S. in physics from the University of Notre Dame, where he now also holds an appointment as a Visiting Scholar in the Reilly Center for Science, Technology, and Values. General Latiff is a graduate of the National Security Fellows Program at Harvard University's John F. Kennedy School of Government. He is a recipient of the National Intelligence Distinguished Service Medal and the Air Force Distinguished Service Medal.

Terry P. Lewis is the Buena Park site executive and manager of ACT/Engineering Services Group with the Raytheon Company. There, his areas of expertise include command, control, communications, and information systems; digitized battlespace systems; communications and transmission security in military tactical systems; wireless network security; and network management authentication techniques for robust security architecture. In addition, Dr. Lewis has developed anti-tampering technologies to prevent or reduce the ability of potential aggressors to reverse-engineer critical U.S. technologies. He is a former Raytheon Engineering Scholar and Fellow and received the Most Promising Engineer of the Year Award, conferred at the 2002 Black Engineer of the Year Award Conference. Dr. Lewis served as a member of the National Research Council's Committee on Distributed Remote Sensing for Naval Undersea Warfare.

Michael A. Longoria currently serves as an adjunct senior analyst at the RAND Corporation, providing operations and analytical support on force modernization research involving intelligence, surveillance, reconnaissance, command, and control systems; close air support; combat search and rescue; special airlift; and special operations. He also trains and helps educate graduate and doctoral cultural anthropology and sociology candidates in the Human Terrain System for the U.S. Army at St. Mary's University, Leavenworth, Kansas, as an independent contractor. He retired from the U.S. Air Force in 2009 as a brigadier general, having held management and analytical positions as well as commanding at all tactical levels within the Air Force and in combat. He served as the director for Democracy Programs at the National Security Council staff, White House, as the Special Assistant to the President helping direct the humanitarian resettlement and security operations for more than 75,000 Cuban and Haitian refugees during the Clinton administration. Additionally, he directed the antiterrorism effort for the Department of Defense (DoD) in the Office of the Secretary of Defense. As a general officer, he represented the DoD in Lyon, France, at Interpol for the arrest and capture of 35 of the top 50 known terror suspects as a result of combat operations in Afghanistan and Iraq. He has extensive combat experience in Panama, Bosnia, Kosovo, Afghanistan, Somalia, Iraq, and Saudi Arabia, but he has also conducted rescue and humanitarian operations in the Philippines, Honduras, Kenya, Liberia, the U.S. Virgin Islands, and the U.S. Gulf Coast after Hurricane Katrina. In his career, he has been assigned to and in support of the following Joint and/or other-service units: 5th and 7th Special Forces Groups, 75th Ranger Regiment, 82nd Airborne Division, 2nd Marine Division, XVIII Airborne Corps, U.S. Army Special Operations Command, and U.S. Third Army; in Operation Iraqi Freedom he commanded all Air Force units assigned to every maneuver division in the U.S. Army. He has served in fellowships at Harvard University and the Congressional Research Service and holds the M.A.A.S. in air and space science, School of Advanced Airpower, Space and Cyber

Studies, Maxwell Air Force Base, Alabama, as well as an M.A. in national security, Naval War College, Newport, Rhode Island.

Paul F. McManamon is the president of Exciting Technology, LLC. He also works half-time as the technical director of the Ladar and Optical Communications Institute at the University of Dayton. Until May 2008, he was chief scientist for the Air Force Research Laboratory's (AFRL's) Sensors Directorate, which consists of approximately 1,250 people responsible for developing new sensor technology for the Air Force. Dr. McManamon was responsible for the technical portfolio of the Sensors Directorate, including radio-frequency sensors and countermeasures, electro-optical (EO) sensors and countermeasures, and automatic object recognition. He has developed multidiscriminate EO sensors, including multifunction laser radar, novel EO countermeasure systems, and optical phased-array beam steering. Dr. McManamon has participated in three Air Force Scientific Advisory Board summer studies: New World Vistas (1995), A Roadmap for a 21st Century Aerospace Force (1998), and Sensors for Difficult Targets (2001). He was instrumental in the development of laser flash imaging, initiating the ERASER program as a method to enhance EO target-recognition range by a factor of 4 or 5. Dr. McManamon is widely recognized in the electro-optical community. He was the 2006 president of SPIE, the international society for optics and photonics. He was on the SPIE board of directors for 7 years and on the SPIE Executive Committee from 2003 through 2007. He serves on the executive committee for the Military Sensing Symposia (MSS) and is a fellow of SPIE, the Institute of Electrical and Electronics Engineers, the Optical Society of America (OSA), AFRL, and MSS.

Matt L. Mleziva is currently the president of Wildwood Strategic Concepts, LLC, a strategic management company in Westford, Massachusetts. Mr. Mleziva has led Joint teams for the Office of the Secretary of Defense that developed recommendations projected to save millions of dollars annually. He guided U.S. Air Force Networked Tactical Communications efforts into a single Joint program with the U.S. Navy. Mr. Mleziva has a proven track record of achieving cost, schedule, and performance goals across organizations covering a wide range of information system technologies for a diverse customer base. He acquired space, air, and electronic systems for the Department of Defense, the U.S. government, and foreign nations. Mr. Mleziva has a demonstrated capability to utilize emerging information technology and promote commonality and interoperability in combat systems. He developed an ultra-streamlined acquisition strategy in response to urgent Air Force operational needs. Mr. Mleziva is the recipient of several awards, including the Presidential Meritorious Executive Rank Award and the Air Force Outstanding Civilian Career Service Award. He holds a post master's degree in electrical engineering and an M.S. in electrical engineering from the Massachusetts Institute of Technology.

Gerald F. Perryman, Jr., is currently a private consultant. Previously he had been the director of strategic pursuits, Defense and Civil Mission Solutions, for Raytheon Intelligence and Information Systems. Major General Perryman oversaw and coordinated strategies and the development of new business opportunities and pursuits related to integrated tactical intelligence and information systems. He was appointed to that position in February 2006 after having led the Raytheon Intelligence, Surveillance, and Reconnaissance (ISR) Strategic Business Area since joining the company in November 2002. Before joining Raytheon, Maj Gen Perryman served as commander of the Aerospace Command and Control and ISR Center at Langley Air Force Base, Virginia. Earlier, he commanded the 14th Air Force, which encompasses all U.S. Air Force space operations forces worldwide. He received his M.S. in business administration from the University of North Dakota.

Jonathan M. Smith is the Olga and Alberico Pompa Professor of Engineering and Applied Science at the University of Pennsylvania. From 2004 to 2006, he was a program manager at the Defense Advanced Research Projects Agency, for which he received the Office of the Secretary of Defense (OSD) Medal for Exceptional Public Service in 2006. He was elected a fellow of the Institute of Electrical and Electronics Engineers in 2001 for "contributions to the technology of high-speed networking." He was previously at Bell Telephone Laboratories and Bellcore, which he joined at the AT&T divestiture. His current research interests range from programmable network infrastructures and cognitive radios to architectures for computer-augmented immune response. Dr. Smith served on the President's Council of Advisors on Science and Technology Network and Information Technology Technical Advisory Group. He was a member of the National Research Council's (NRC's) Committee on Sensing and Communications Capabilities for Special Operations Forces and is a current member of the NRC Board on Army Science and Technology.

Appendix B

List of Committee Meetings, Presenters, and Participating Organizations

MEETING 1
OCTOBER 6-7, 2011
THE KECK CENTER OF THE NATIONAL ACADEMIES
WASHINGTON, D.C.

Air Force Acquisition
Dr. Steven H. Walker, Deputy Assistant Secretary of the Air Force (Science, Technology, and Engineering)

Initial Session Remarks
Lt Gen Larry D. James, Deputy Chief of Staff for Intelligence, Surveillance, and Reconnaissance, Headquarters, U.S. Air Force

SAF/AQI Presentation to the Committee on Examination of the Air Force Intelligence, Surveillance, and Reconnaissance (ISR) Capability Planning and Analysis (CP&A) Process
Col Gregory Gutterman, Deputy Director, Information Dominance Programs (SAF/AQI)

Air Force ISR: CP&A Overview
Col Brian D. Johnson, Chief, Air Force Plans and Integration Division for the Deputy Chief of Staff, Intelligence, Surveillance and Reconnaissance, Headquarters, U.S. Air Force

The Air Force ISR Flight Plan: Origin, Rationale, and Process
Lt Gen David Deptula (USAF, Ret.), Chief Executive Officer and Managing Director, Mav6, LLC

Core Function Lead Integrator Concept
Col Scot Gere, Global Integrated ISR [GIISR] Core Function Team Chief, Air Combat Command, Langley Air Force Base, Virginia

GIISR Core Function Master Plan Overview
Mr. Mike Kennedy, Planning Team Lead, GIISR Core Function Master Plan, Langley Air Force Base, Virginia

Office of the Under Secretary of Defense (Intelligence)(OUSD[I]) Overview
Col Tony Lombardo, Deputy Director, ISR Programs, Agency Acquisition Oversight, OUSD(I)

AF/A9 Presentation to the Committee
Mr. Kevin Williams, Principal Deputy Director, AF/A9, Headquarters, U.S. Air Force

Office of the Assistant Secretary of the Air Force for Acquisition (SAF/AQS) Presentation to the Committee
Maj Gen John E. Hyten, Director, Space Programs, SAF/AQS

MEETING 2
NOVEMBER 9-10, 2011
THE KECK CENTER OF THE NATIONAL ACADEMIES
WASHINGTON, D.C.

Warfighting Integration
Mr. Brian Burns, Deputy Director, Warfighter Systems Integration, Office of Information Dominance and Chief Information Officer, Office of the Secretary of the Air Force

Air-Sea Battle and the Air Force ISR CP&A Process
Maj Gen Rick Devereaux, Director of Operational Planning, Policy and Strategy, Headquarters, U.S. Air Force

Operationally Responsive Space (ORS) Office Capability Planning and Analysis Processes
Dr. Peter Wegner, Director, ORS Office, Kirtland Air Force Base, New Mexico

Committee on the Air Force ISR Capability and Planning Process
LTG Richard Zahner, Deputy Chief of Staff, G-2, Headquarters, U.S. Army

Evolving Air Force Space Command (AFSPC) ISR Corporate Planning
Col Christina Morris, HQ AFSPC Deputy Director, ISR, Lead, BATI Capability Team

Capture the Past, Build the Future: Capability Planning and Analysis
Brig Gen Dwyer L. Dennis, Director, Intelligence and Requirements Directorate, Headquarters, Air Force Materiel Command, Wright-Patterson Air Force Base, Ohio

Secretary of the Air Force ISR Review Results
Mr. Mark "Tap" Tapper, Special Adviser to the Deputy Chief of Staff for Intelligence, Surveillance and Reconnaissance, Headquarters, U.S. Air Force

MEETING 3
DECEMBER 7-8, 2011
THE KECK CENTER OF THE NATIONAL ACADEMIES
WASHINGTON, D.C.

Congressional Perspectives
Senate Select Committee on Intelligence:
- Ms. Amy Hopkins, Professional Staff Member
- Ms. Peggy Evans, Budget Director
- Mr. Jim Wolfe, Professional Staff Member

House Permanent Select Committee on Intelligence:
Ms. Brooke Eisele, Professional Staff Member

Capabilities-Based Planning and Analysis at the National Geospatial-Intelligence Agency (NGA)
Mr. Winston Beauchamp, Technical Executive, NGA

Space ISR Planning and Forecasting
Mr. Doug Loverro, Executive Director, Space and Missile Systems Center, Air Force Space Command

Missile Defense Agency (MDA) Capability Planning and Analysis
Mr. Rich Ritter, Program Executive Officer for C4ISR, MDA

Roles and Responsibilities of the Capabilities Portfolio Manager
Mr. Kenneth E. Bray, Technical Adviser, Intelligence Programs and Budget, to the Director of Intelligence, Surveillance and Reconnaissance Programs, Deputy Chief of Staff for Intelligence, Surveillance and Reconnaissance, Headquarters, U.S. Air Force

National Reconnaissance Office (NRO) Presentation to the Committee
Maj Gen Susan K. Mashiko, Deputy Director, NRO

Systems and Resource Analyses Organization and Process
Mr. David Svetz, Deputy Assistant Director of National Intelligence for Systems and Resource Analyses, Office of the Director for National Intelligence

Modeling of ISR Environments and Processing, Exploitation, and Dissemination
Dr. Lance Menthe, Physical Scientist, RAND Corporation

Cyberspace Intelligence, Surveillance, and Reconnaissance
Col Tom French, Chief of Plans, Strategies AFSPC/A2

ISR for the Future
Maj Gen Blair E. Hansen, Deputy Commander, Joint Functional Component Command for Intelligence Surveillance and Reconnaissance; and Deputy Director of the Defense Intelligence Agency for Collection Management

MEETING 4
JANUARY 5-6, 2012
THE KECK CENTER OF THE NATIONAL ACADEMIES
WASHINGTON, D.C.

Capabilities-Based Portfolio Management: Methods, Processes, and Tools
Mr. Scott Gooch, Principal, Booz Allen Hamilton
Mr. Christopher Anderson, Lead Associate, Booz Allen Hamilton

Google's Approach to Capability Planning and Analysis
Ms. Michele Weslander-Quade, Chief Technology Officer, Google, Inc.

An Enterprise Approach to Capability-Based Analysis: Best Practices, Tools, and Results
Mr. Doug Owens, Manager, Enterprise Analysis, Defense Business Unit, TASC.

CP&A Study Sponsor Discussion with the Committee
Lt Gen Larry James, Deputy Chief of Staff for ISR, Headquarters, U.S. Air Force

RadiantBlue Modeling and Simulation Capability
Mr. Larry Shand, President, RadiantBlue, Inc.
Mr. Phil Eichensehr, Vice President, RadiantBlue, Inc.
Mr. Patrick O'Neill, Division Manager, RadiantBlue, Inc.

U.S. Cyber Command's Approach to Capability Planning and Analysis for the Cyber Domain
Mr. Everett (Rusty) Rollins, Deputy Director of the Joint Intelligence Operations Center, U.S. Cyber Command, Directorate of Intelligence (J2)

Layered ISR Architecture Analysis
Mr. Kurt Dittmer, Advanced Programs and Technology, Advanced Projects Director, Northrop Grumman Corporation

Metropolitan Police Department's Strategy for Interoperability
Mr. Tom Wilkins, Executive Director, Intelligence Fusion Division, Metropolitan Police Department, Washington, D.C.

MEETING 5
FEBRUARY 18, 2012
AIR FORCE ISR AGENCY (AFISRA)
LACKLAND AIR FORCE BASE, TEXAS

AFISRA Mission Brief
SSgt Westbrooks, AFISRA/CCX

AFISRA and the CP&A Process
Mr. Donald Schiber, AFISRA/A5R

AF DCGS Enterprise
Col Michael Shields, AFISRA/A5W

AFISRA Corporate Process
Mr. Kurt Eversole, AFISRA/A8P

24th Air Force (24AF) Mission Brief
Col Matthew "Sunshine" Baker, 24AF/A2

ISR Support Requirements
Maj Karin Reynolds, 24AF/A2X

24AF Cyber ISR Requirements Working Group
Lt Col Leonard "Len" Pilhofer, 24AF/A2X/Z

MEETING 6
FEBRUARY 25, 2012
AIR COMBAT COMMAND (ACC)
LANGLEY AIR FORCE BASE, VIRGINIA

Introductory Comments
Lt Gen William J. Rew, Vice Commander, ACC

Air Force GIISR Core Function Lead Integrator Construct
Col Scot Gere, GIISR Core Function Team Chief

Roundtable Discussion
Col Scot Gere, GIISR Core Function Team Chief
Col William Pinter, ACC/A5S
Col Richard Donnelly, ACC/A5P
Col Eric Holdaway, ACC D-A2
Mr. Robert Burgess, ACC/A8X

MEETING 7
FEBRUARY 14-16, 2012
ARNOLD AND MABEL BECKMAN CENTER
IRVINE, CALIFORNIA

The National Security Agency's Approach to Capability Planning and Analysis
Mr. Wayne Landry, Deputy Director, ISR PMO

Writing Meeting

MEETING 8
FEBRUARY 22, 2012
THE PENTAGON
ARLINGTON, VIRGINIA

The U.S. Navy's Approach to Capability Planning and Analysis
RADM DeWolfe Miller, Director, Intelligence, Surveillance and Reconnaissance, Capabilities Division (Opnav N2/N6f2)

MEETING 9
MARCH 27-29, 2012
ARNOLD AND MABEL BECKMAN CENTER
IRVINE, CALIFORNIA

Writing Meeting

Appendix C

Supplement to Chapter 3: Descriptions of Additional Organizational CP&A Processes and Tools

MISSILE DEFENSE AGENCY

The Missile Defense Agency (MDA) employs a capabilities analysis process for establishing requirements that relies heavily on modeling and simulation.[1] The MDA seeks to engage the services and Combatant Commands (COCOMs) in the process. Exempt from the Joint Capabilities Integration and Development System (JCIDS) requirements process, MDA has implemented the Warfighter Involvement Process (WIP), represented in Figure C-1.

The WIP is designed to align the MDA program with warfighter priorities for missile defense in a streamlined manner. The process starts with the U.S. Strategic Command (USSTRATCOM) soliciting COCOM inputs, which are based on the Defense Planning Guidance, threat developments, and the commander's intent. The MDA consolidates and prioritizes these inputs into a Prioritized Capability List (PCL). The PCL is intended to be written as capability needs, not to include specifics of weapon system solutions. Within the JCIDS process, it is analogous to the Initial Capability Document (ICD).

The MDA receives the PCL and performs an analysis of the baseline Program of Record (PoR) in order to identify gaps in capability. The MDA captures the parameters associated with the Ballistic Missile Defense System (BMDS) elements,

[1]Further information on MDA's testing and capabilities analysis process is available through the MDA web site: "Supporting Efforts: Ballistic Missile Defense Testing." Available at http://www.mda.mil/system/testing.html. Accessed February 28, 2012.

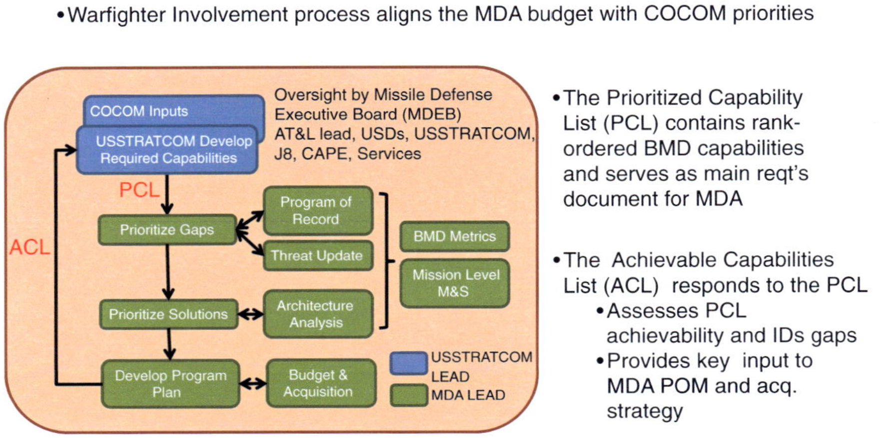

FIGURE C-1 The Missile Defense Agency's (MDA's) Warfighter Involvement Process. NOTE: Acronyms are defined in the list in the front matter. SOURCE: Missile Defense Agency.

as funded in the PoR, in the Element/Component Characterizations for Analysis (E/CCA). The threat picture is also updated with the latest intelligence assessments. The E/CCA and the updated threat picture are the analytic basis for identifying gaps relative to the PCL. Element-level models such as the Aerospace Corporation's System Performance Evaluation Tool (SPET) model for space-system tracking quality are used to generate element capability parameters that are used in BMDS engagement-level modeling (e.g., WILMA). These top-level models are used to generate the BMDS-level metrics, including defended area, raid capacity, probability of engagement success, among others. The MDA architecture team performs architecture assessment and proposes solutions to mitigate the gaps. These potential solutions are assessed for gap closure using the same modeling and analysis tools.

Proposed solutions are prioritized by senior MDA leadership, and the highest-priority solutions are selected for further requirements assessment, design work, and top-level costing. The element-level requirements process is unlike the JCIDS process in that the systems engineering and architecture process flows requirements down from the PCL rather than performance requirements being directly provided by the users. The lead services still provide Doctrine, Organization, Training, Materiel, Leadership and Education, Personnel and Facilities (DOTMLPF) requirements directly. The prioritized solutions are evaluated against the anticipated budgets, and the director-approved Achievable Capabilities List (ACL) is briefed to the Missile Defense Executive Board for final approval. The ACL is then provided to USSTRATCOM as a response to the PCL. The element program modifications and new program starts are approved by Office of the Under Secretary of Defense for

Acquisition, Technology, and Logistics (OUSD[AT&L]). The MDA uses a number of tools that have been developed over time, including the following:

- *The WILMA-Suite* is an end-to-end modeling and analysis suite for BMDS concept exploration used to predict the performance of missile defense architectures. WILMA is a medium-fidelity end-to-end simulation evaluation of missile defense architecture effectiveness, with consideration for battle planning, communications, and integration alternatives. WILMA may also be run in a higher-fidelity mode, used in conjunction with other in-depth element, engineering, environment, phenomenology, and threat models to make detailed assessments of missile defense system performance against adversary ballistic missile attacks.
- *The SPET* models space-sensor architecture performance, including multi-target tracking, scheduling, and handling association errors. SPET provides ground-, air-, and space-infrared (IR)/visible-sensor system performance against missile threat scenarios; the tool provides target signatures, coverage analysis, position and velocity tracking accuracy distributions, and interceptor effectiveness.
- *SYSSIM (System Simulation)* also models space-sensor architecture performance and includes interceptor flyouts (surface- or air-based). SYSSIM models the interceptor's kinematic flyout or reach and also the probability of kill (Pk) of the engagement. SYSSIM provides engagement time lines and Pk as a function of space-/air-/ground-sensor tracking accuracy.

Observations and Attributes

Although the MDA process described above is specific to MDA requirements, it necessarily considers intelligence, surveillance, and reconnaissance (ISR) architectural assets (particularly communications and PCPAD/TCPED [planning and direction, collection, processing and exploitation, analysis and production, and dissemination/tasking, collecting, processing, exploiting, and disseminating]) in order to deliver required capability. The MDA is exempt from the JCIDS process, which streamlines decision time lines somewhat and has fostered the development of a set of analysis tools that continue to evolve.

OPERATIONALLY RESPONSIVE SPACE OFFICE

The Operationally Responsive Space (ORS) Office was established (1) to contribute to the development of low-cost, rapid-reaction payloads, buses, spacelift, and launch-control capabilities in order to fulfill Joint military operational requirements for on-demand support and reconstitution; and (2) to coordinate and

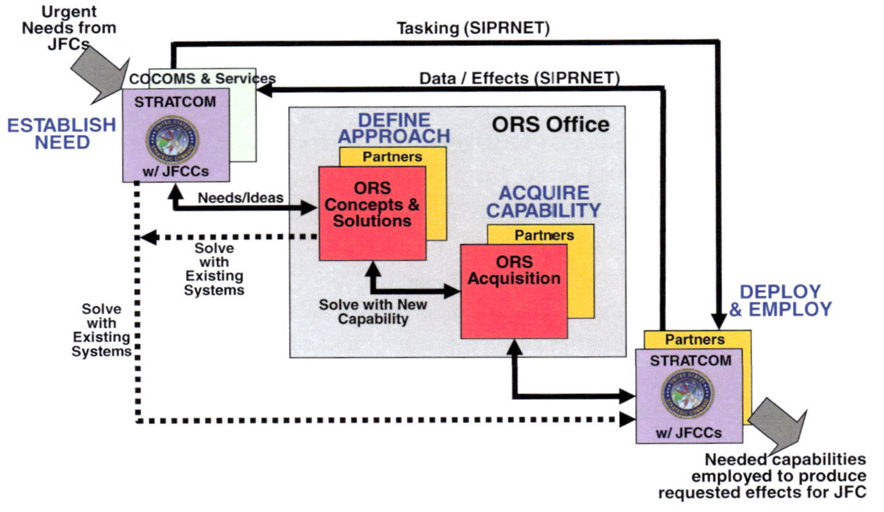

FIGURE C-2 The Operationally Responsive Space (ORS) Office process provides three tiers of response. SOURCE: ORS Office.

execute operationally responsive space efforts across the Department of Defense (DoD) with respect to planning, acquisition, and operations.[2] The ORS Office receives its tasking and requirements from the Commander, U.S. Strategic Command (CDRUSSTRATCOM) and has a streamlined reporting structure directly to the DoD Executive Agent (EA) for Space.[3] CDRUSSTRATCOM developed an initial concept of operations (CONOPS) for the ORS Office that specifies a number of warfighting effects expected by achieving an ORS end-state. Those effects are the abilities to reconstitute lost capabilities, to augment or surge existing capabilities, to fill unanticipated gaps in capabilities, to exploit new technical and operational innovations, to respond to unforeseen or episodic events, and to enhance survivability and deterrence. Additionally, the CONOPS provides a requirement to develop a three-tiered approach for delivering capability, as shown in Figure C-2.[4]

The Tier 1 requirement is the ability to employ existing space assets (on orbit) in minutes to hours in order to meet an unforeseen need. The Tier 2 requirement

---

[2]Information on the mission of the ORS Office is available at http://ors.csd.disa.mil/mission/index.html. Accessed February 28, 2012.

[3]Col Ken McLaughlin, Director, ORS Office. 2007. "Operationally Responsive Space Office." Slides dated July 2007. Available at http://www.responsivespace.com/ors/reference/McLaughlin.pdf. Accessed February 28, 2012.

[4]ORS Program Office. More information on the three tiers of response is available at http://ors.csd.disa.mil/tier-1/index.html. Accessed February 28, 2012.

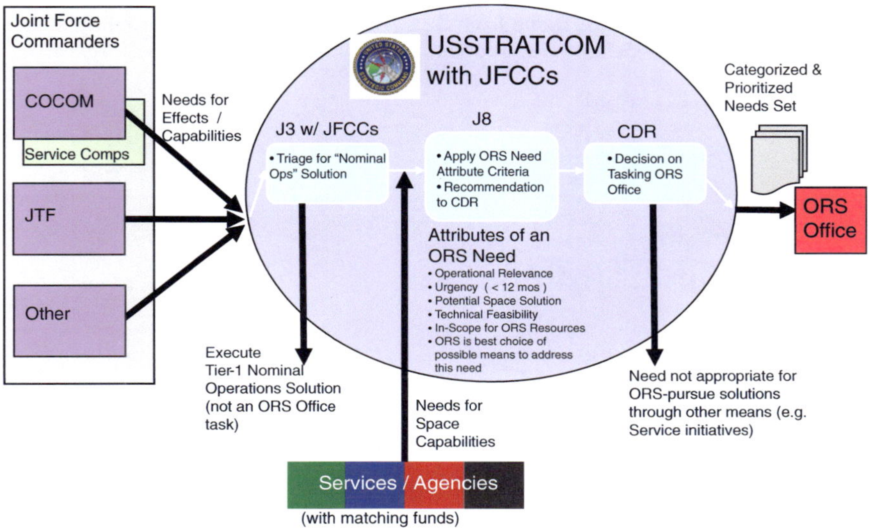

FIGURE C-3 Step 1 of the Operationally Responsive Space (ORS) Office process: Establish need. SOURCE: ORS Office.

is the ability to launch and/or deploy "field ready" assets in days to weeks; this implies an "on call" capability. The Tier 3 requirement is the ability to rapidly develop and transition to delivery a new or modified capability in months (not years). CDRUSSTRATCOM has developed a process of seeking urgent needs from combatant commanders and then prioritizing these needs and assigning tasking to the ORS Office, which develops alternatives to meeting the urgent need. This process is depicted in Figure C-3.

After establishing the need, the ORS Office then defines the approach to meeting the identified need, which involves collaboration and iteration, as shown in Figure C-4. Here, requirements are defined, prioritized, and validated; then a solutions analysis process follows, which is ultimately presented to CDRUSSTRATCOM and the EA for Space.

In addition to using subject-matter experts, the ORS Office uses several tools that fall into five categories, described briefly below. Of note, the tools help implement the process more effectively and serve as aids to inform decisions.

Architecting Tools

The ORS Office response to urgent needs is really an analysis of alternatives in a very compressed time line (typically 45 days from receipt of task from CDRUSSTRATCOM to the delivery of recommended courses of action to both

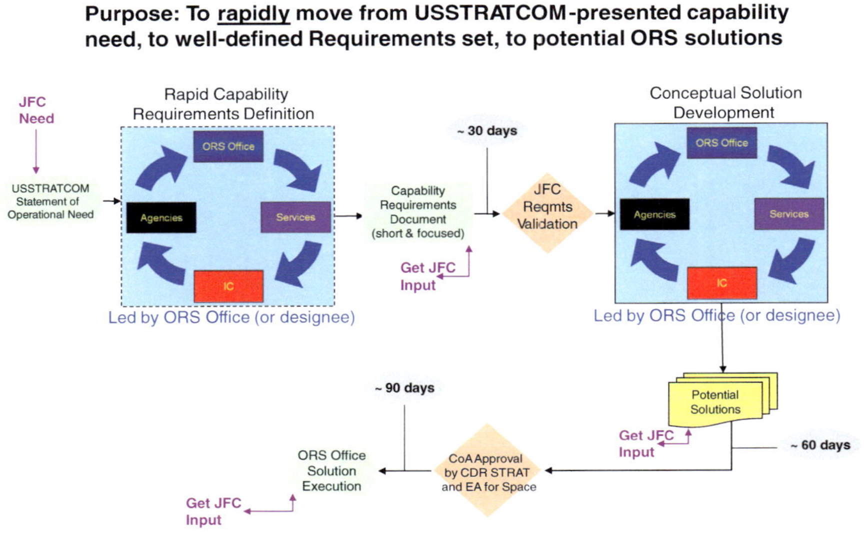

FIGURE C-4 Step 2 of the U.S. Strategic Command (USSTRATCOM) process: Define approach. SOURCE: ORS Office.

CDRUSSTRATCOM and DoD EA for Space). ORS personnel typically use a variation of the Aerospace Systems Architecting Process (ASAP). The key is getting the purpose analysis and problem framing done right up-front, by converting the Statement of Need from the requesting Joint Force Commander or Joint Functional Component Commander into the Capabilities Requirements Document (CRD). The CRD contains a list of prioritized evaluation criteria from the user that help evaluate courses of actions.

Interacting/Communicating Tools

When the requirements and concepts/solutions teams are working and having teleconferences, it is important to have the proper communications capabilities in place. The ORS Office developed a "hyper community of practice (CoP)" in 2008 as a mechanism to capture and share data among all team members. The hyper-CoP is a "SharePoint"-based tool that allows team members to have joint and immediate access to all products developed by the teams.

Modeling, Simulation, and Analysis Tools

The fast pace of the response often precludes detailed modeling, simulation, and analysis (MSA) work in the concept-formulation phase. However, the office used some Aerospace imagery simulation tools in preparing for the start of the ORS-1 development process and used campaign models (e.g., System Effectiveness Analysis Simulation [SEAS]) when completing the Joint Military Utility Analyses after the launch of capabilities such as ORS-1, TacSat-3, and TacSat-4.

Visualization Tools

The ORS Office uses three visualization tools: Satellite Tool Kit (STK)®; Satellite Orbit Analysis Program (SOAP); and the Advanced Geospatial Intelligence Tool Kit (AGI/TK). These tools facilitate the comparison of orbits' "best fit" in meeting a need.

Decision Support Tools

The ORS Office uses Expert Choice® to assist in pair-wise comparisons and decision support when the concepts/solutions team is finalizing its recommendations. (The tool Expert Choice is provided through the company Decision Lens.) It is a widely used tool, often employed to assist source selection teams in documenting choices made during competitive source selections. Decision Lens has also developed enterprise architecture/knowledge-management tools, without great success.

Observation and Attributes

The ORS process is streamlined in that it takes typically well-defined, "urgent" requirements from Joint Force Commanders, develops alternative solutions, and presents recommendations directly to CDRUSSTRATCOM and the DoD EA for Space, substantially reducing the time line for decisions. The process takes into consideration ISR capabilities across the enterprise. In general, the process does not directly address the emergent cyber threats or considerations.

NATIONAL RECONNAISSANCE OFFICE

The National Reconnaissance Office (NRO) collaborates with its mission partners and capability managers across the intelligence community (IC) and the Department of Defense, as depicted in Figure C-5. In many cases, the NRO both acquires and operates systems.

The NRO uses a process, led by its Chief System Engineering Office, to perform system- and architectural-level analysis to inform investment decisions, as shown in Figure C-6. The process uses a top-down approach through which strategic guidance and mission needs are decomposed, followed by an assessment of current capabilities in order to identify shortfalls and gaps. Next is an iterative process

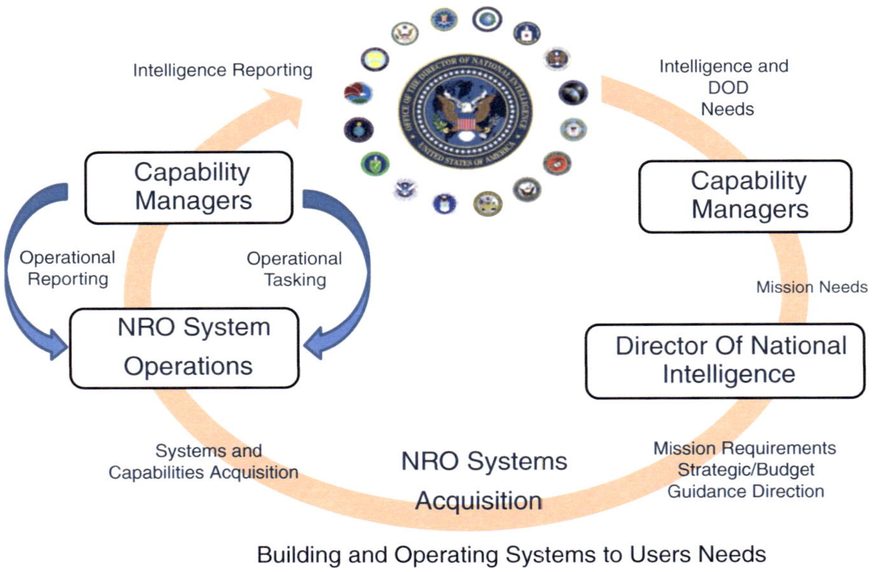

FIGURE C-5 The National Reconnaissance Office's (NRO's) capability evaluation and delivery process. SOURCE: Maj Gen Susan Mashiko, Deputy Director, National Reconnaissance Office. "Briefing to the Committee on Examination of the Air Force ISR CP&A Process." Presentation to the committee, December 8, 2011.

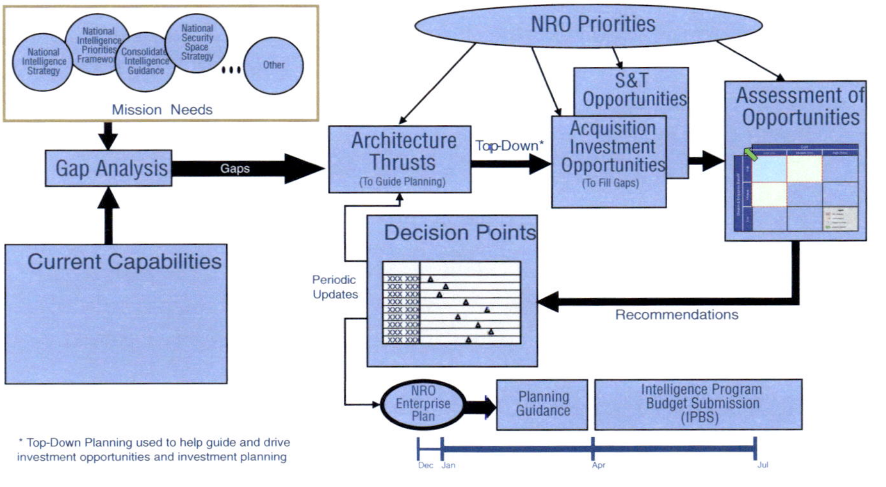

FIGURE C-6 The National Reconnaissance Office's (NRO's) integrated architecture and investment planning process. SOURCE: National Reconnaissance Office.

that evaluates alternative solutions based on NRO and operational priorities for informing potential investment decisions. The NRO uses a variety of tools, both commercial off-the-shelf and custom.

RAND CORPORATION

The RAND Corporation uses two main systems engineering tools: the Systems and CONOPS Operational Effectiveness Model (SCOPEM) developed by RAND and the SEAS Modeling Environment maintained and developed by ExoAnalytic Solutions for the Air Force Space Command, Space and Missile Systems Center, Directorate of Development Planning (SMC/XR). SEAS is part of the Air Force Standard Analysis Toolkit and the Air Force Space Command Modeling and Simulation Toolkit (see Figure C-7).

The SCOPEM tool is designed to examine tasking, collection, and targeting decisions within the SEAS Modeling Environment and offers flexible output for the development of measures of performance and/or effectiveness (such as target quality like the National Imagery Interpretability Rating Scale [NIIRS] rating or signal-to-noise ratio and predetermined metrics such as the number of signals intelligence [SIGINT] cues successfully prosecuted in an hour). See Figures C-8 and C-9.

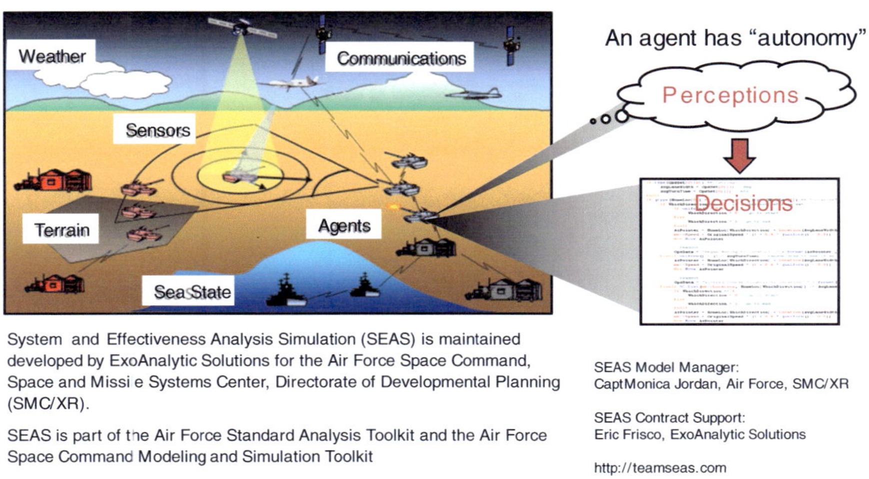

FIGURE C-7 The System Effectiveness Analysis Simulation (SEAS) tool is part of the Air Force Space Command Modeling and Simulation Toolkit used by the RAND Corporation. SOURCE: Lance Menthe, Physical Scientist, RAND Corporation. "The Systems and CONOPS Operational Effectiveness Model (SCOPEM) and Systems Effectiveness Analysis Simulation (SEAS) Modeling Environment." Presentation to the committee, December 7, 2011.

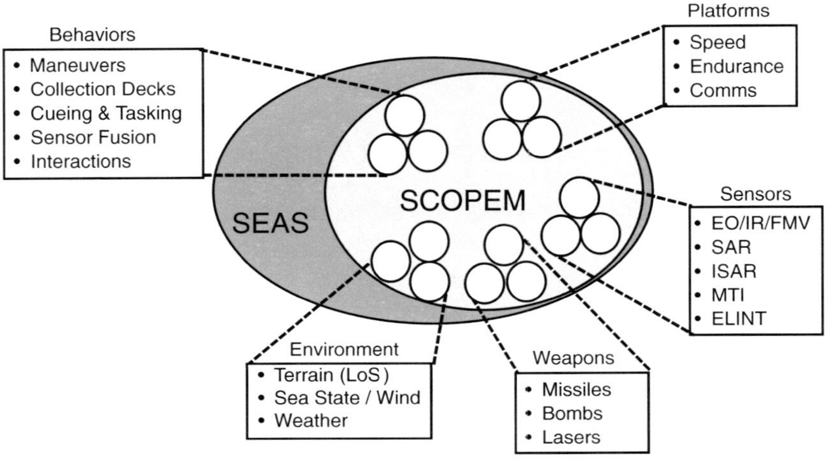

FIGURE C-8 RAND Corporation tool written for the System Effectiveness Analysis Simulation (SEAS) modeling environment. NOTE: The Systems and CONOPS Operational Effectiveness Model (SCOPEM) was formerly known as the Collections Operations Model. SOURCE: Lance Menthe, Physical Scientist, RAND Corporation. "The Systems and CONOPS Operational Effectiveness Model (SCOPEM) and Systems Effectiveness Analysis Simulation (SEAS) Modeling Environment." Presentation to the committee, December 7, 2011.

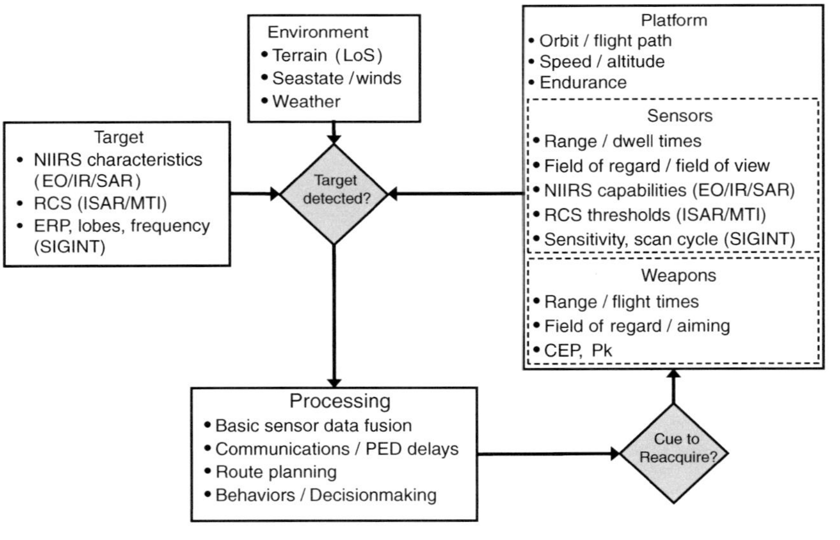

FIGURE C-9 The RAND tool Systems and CONOPS Operational Effectiveness Model (SCOPEM) is designed to examine tasking, collection, and targeting decisions. SOURCE: Lance Menthe, Physical Scientist, RAND Corporation. "The Systems and CONOPS Operational Effectiveness Model (SCOPEM) and Systems Effectiveness Analysis Simulation (SEAS) Modeling Environment." Presentation to the committee, December 7, 2011.

The utility of this simple tool was illustrated by a series of sensor, weapon, and environment models such as generic sensor models, electro-optical (EO)/ infrared (IR)/synthetic aperture radar (SAR) and electronics intelligence (ELINT) imaging performance estimates, ground moving target indicator (GMTI) tracking models, general weapons models, a missile model, a directed-energy weapon (laser) model, and environmental effects. Four recent examples of SCOPEM/SEAS evaluation scenarios include (1) Global Hawk sensor performance in various maritime operations: MCO-1, MCO-2, and maritime interdiction; (2) air-based and space-based maritime domain awareness in the Mediterranean region; (3) evaluation of F-22 sensor capabilities; and (4) finding, identifying, and tracking of vehicles, using different classes of remotely piloted aircraft, under varied environmental conditions. In summary, SCOPEM is a very accessible, text input-, personal computer/ Windows-based tool by which complex behaviors are built out of simple behaviors and interactions. Like the more complex Multi-Resolution Analysis (MRA) processes advocated by TASC and RadiantBlue, it is a repeatable, transparent, and understandable tool.

U.S. CYBER COMMAND

The U.S. Cyber Command (USCYBERCOM) is a subunified command subordinate to USSTRATCOM. USCYBERCOM centralizes command of cyberspace operations, organizes existing cyber resources, and synchronizes the defense of U.S. military networks.[5] USCYBERCOM plans, coordinates, integrates, synchronizes, and conducts activities to direct the operations and defense of specified DoD information networks; and to prepare to and, when directed, conduct full-spectrum military cyberspace operations in order to enable actions in all domains, to ensure U.S./Allied freedom of action in cyberspace, and to deny the same to adversaries. The command is charged with pulling together existing cyberspace resources, creating synergy and synchronizing warfighting effects to defend the information security environment.[6] USCYBERCOM is tasked with centralizing the command of cyberspace operations, strengthening DoD cyberspace capabilities, and integrating and bolstering DoD's cyber expertise. As depicted in Figure C-10, USCYBERCOM requirements range from strategic to operational to tactical requirements, with significant emphasis in the operational to tactical arena owing to the dynamic cyber/ network environment and very short time lines for action and reaction.[7]

---

[5]U.S. Air Force (USAF). 2010. "Cyberspace Operations. Air Force Doctrine Document 3-12." Available at http://www.fas.org/irp/DoDdir/usaf/afdd3-12.pdf. Accessed February 27, 2012.

[6]Ibid.

[7]Everett (Rusty) Rollins, Deputy Director of the Joint Intelligence Operations Center, USCYBERCOM, Directorate of Intelligence (J2). "USCYBERCOM's Approach to Capability Planning and Analysis for the Cyber Domain." Presentation to the committee, January 5, 2012.

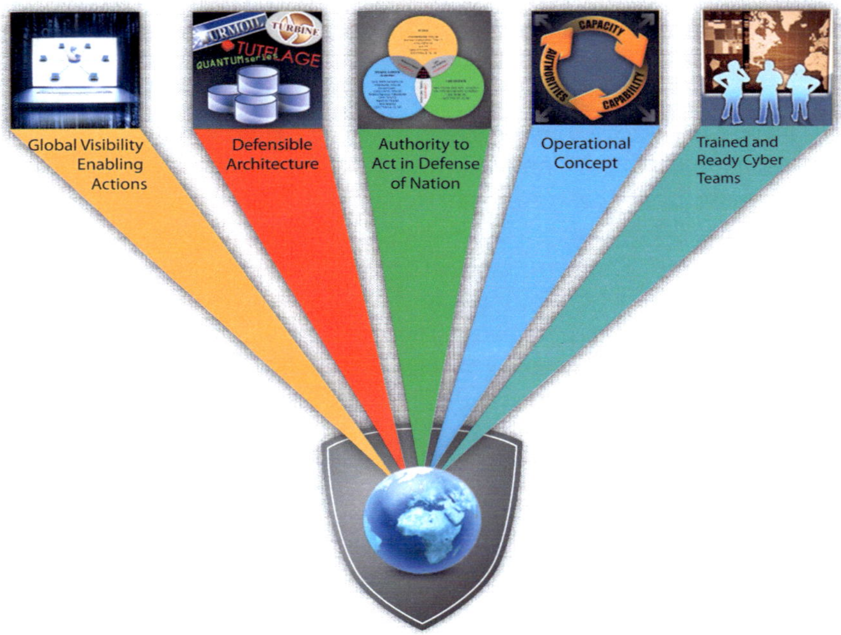

FIGURE C-10 U.S. Cyber Command (USCYBERCOM) requirements for operating in cyberspace. SOURCE: Everett (Rusty) Rollins, Deputy Director of the Joint Intelligence Operations Center, USCYBERCOM, Directorate of Intelligence (J2). "USCYBERCOM's Approach to Capability Planning and Analysis for the Cyber Domain." Presentation to the committee, January 5, 2012.

Additionally, USCYBERCOM responsibilities and authorities cross multiple organizations and commands as well as Title 10 (DoD) and Title 50 (IC) to support both offensive and defensive activities.[8,9]

GOOGLE, INC.

Google was invited to share its capabilities planning process as a potential example of a successful commercial enterprise in a very dynamic, fast-paced, global arena. Interestingly, Google did not present a "process" chart; its investment decisions are based on an evaluation by the company leadership of cost, return

[8]Ibid.

[9]The military services and the IC have specific roles and responsibilities, as defined by Title 10 and Title 50 of the U.S. Code, respectively. For additional information on Title 10, see http://uscode.house.gov/download/title_10.shtml. For additional information on Title 50, see http://uscode.house.gov/download/title_50.shtml. Accessed March 21, 2012.

on investment, and probability of success.[10] Google is focused on responding to customer needs and on continuing to grow a collaborative environment, and it is perhaps more risk-tolerant than many other organizations.[11] Google's decision processes are expeditious, in keeping with the dynamics of the fast-evolving cloud network and social media arena. Once approved, if an initiative's measured results fall short of expectations, the activity is generally curtailed, with lessons learned being documented and shared across the company; hence the phrase "Fail fast/ Fail smartly."[12,13]

Observations and Attributes

The capabilities planning process at Google is expeditious owing to the company's "flat" organizational structure, with relatively few decision makers. The culture and operational concept encourage risk taking and are not overly encumbered with deep layers of analysis and recursive iterations. Customer utilization metrics are used effectively to evaluate and inform investment decisions.

NORTHROP GRUMMAN CORPORATION

Northrop Grumman Corporation (NGC) provides system engineering analysis to the U.S. government and consulting services to the ISR community.[14] NGC presented several examples of MSA tools that were attributed to work performed by TASC prior to the spin-off of TASC from NGC. The physics-based tool for rapid strategic analysis to guide investment portfolio analysis was identified as a Layered ISR Architecture Analysis (see Figure C-11). This layered analysis process is used to conduct rapid, system-of-systems capabilities-based analysis of air- and space-based ISR systems to identify effective force mix options for DoD and IC needs.[15] NGC-described attributes of this process include agility, visibility, flexibility and

[10]Michele Weslander-Quade, Chief Technology Officer, Google, Inc. "Google's Approach to Capability Planning and Analysis." Remarks to the committee, January 5, 2012.

[11]More information on the Google mission and philosophy is available at http://www.google.com/about/company/tenthings.html. Accessed February 28, 2012.

[12]Michele Weslander-Quade, Chief Technology Officer, Google, Inc. "Google's Approach to Capability Planning and Analysis." Remarks to the committee, January 5, 2012.

[13]No specific analytic or decision support tools were presented in Google's remarks to the committee.

[14]More information on Northrup Grumman Corporation system engineering and analysis projects is available at http://www.northropgrumman.com/about_us/index.html. Accessed February 28, 2012.

[15]More information on Northrop Grumman's C4ISR products and programs is available at http://www.northropgrumman.com/isr/index.html. Accessed February 28, 2012.

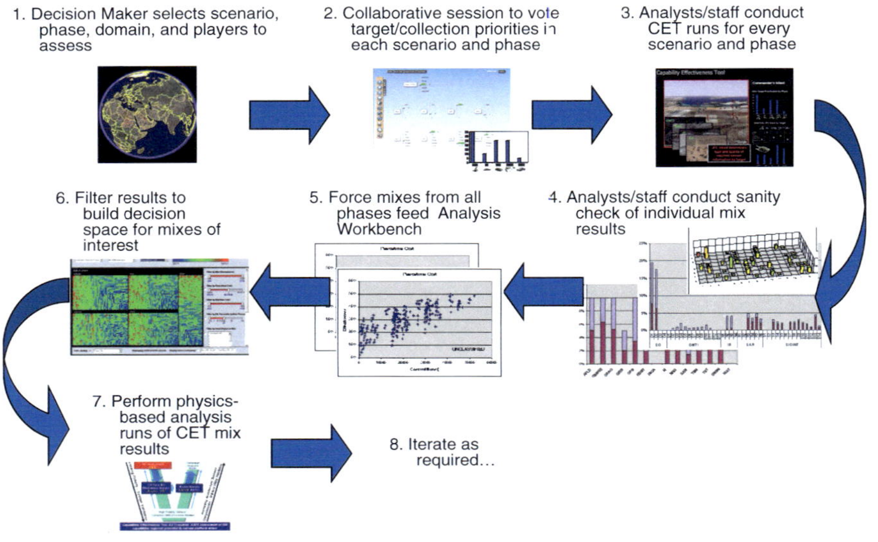

FIGURE C-11 Northrop Grumman Corporation's (NGC's) layered intelligence, surveillance, and reconnaissance (ISR) analysis process. SOURCE: Kurt Dittmer, Advanced Programs and Technology, Advanced Projects Director, Northrop Grumman Corporation. "Layered ISR Architecture Analysis, Collection Capability Discussion." Presentation to the committee, January 5, 2012. Copyrighted material. Used with permission from Northrop Grumman Corporation.

maturity.[16] The process engages the decision maker early and encourages iterative collaboration among operators and analysts for developing an understanding of system and capability trade-offs.

NGC's layered ISR process highlighted a top-level ISR force mix assessment tool—the Layered ISR Capabilities Effectiveness Tool (CET)—created by an NGC team of analysts in order to facilitate Long-Range Strategic Plan trade-offs for ISR systems following the 2006 Quadrennial Defense Review (QDR).[17] This tool was licensed to USSTRATCOM J81 for a proof of principle in 2008 and has been used for studies through a U.S. Joint Forces Command (USJFCOM) ISR Analysis, Integrated Demonstrations and Experimentation Cooperative Research and Development Agreement (CRADA) and is currently being used in support of the Operationally Responsive Space Office and the Missile Defense Agency for capabil-

[16]Kurt Dittmer, Advanced Programs and Technology, Advanced Projects Director, Northrop Grumman Corporation. "Layered ISR Architecture Analysis, Collection Capability Discussion." Presentation to the committee, January 5, 2012.

[17]NGC. 2012. "Layered Intelligence, Surveillance, Reconnaissance (ISR) Architecture Analysis." Point paper. Written communication to the committee.

TABLE C-1 Description of the Layered ISR Capabilities Effectiveness Tool (CET), with Product and Benefits, Employed by the Northrop Grumman Corporation

Tool Name	Category	Tool Product	Benefits
CET: Layered ISR Capabilities Effectiveness Tool	Architecture Analysis	Simulation	Top-level ISR force mix assessment tool to guide physics-based analysis. Orders of magnitude faster than current simulation tools. Can be used with many physics-based tools (e.g., STK, EADSIM, BLUESIM).

NOTE: For a more complete list of tools used by both government and industry, see the full text of Appendix C.
SOURCE: Northrop Grumman Corporation. 2012. Written communication. Response to inquiry from the committee.

ity trade-offs.[18] Table C-1 describes the tool employed by NGC and its products and benefits.

The CET is an ISR tool when considering all costs for a specific force mix. Typical inputs from Air Force stakeholders include (1) theater and area of operations; (2) mission domains and phases of war; (3) Commander's intent (type and quality of required sensor information by target) and target prioritizations; (4) Intel-developed target collection requirements; (5) ISR platforms and satellite constellations, including system performance and sensor capabilities; and (6) system costs. Typical output of the layered ISR tool suite includes (1) all viable force mix solutions; (2) system-of-systems domain, phase, and scenario effectiveness assessment; (3) domain, phase, and scenario capability gaps; (4) system-of-systems costs-effectiveness assessment for life cycle and operations; and (5) optimized orbit locations based on unique system-of-systems solution and platform CONOPS.

In summary, the NGC physics-based analysis layered ISR tool suite was created to improve decision making. It links the commander's intent (information desired) to results, decisions, and, most significantly, costs. The tool suite has been validated by USSTRATCOMJ-81 through USJFCOM CRADA and is currently being used to support Air Force ORS trade-offs.

[18]Ibid.